The Musical Art of Synthesis

New synths with unique features and layers of complexity are released frequently, with hundreds of different synths currently available in the marketplace. How do you know which ones to use and how do you get the most out of the ones you already own? *The Musical Art of Synthesis* presents synthesizer programming with a specific focus on synthesis as a musical tool. Through its innovative design, this title offers an applied approach by providing a breakdown of synthesis methods by type, the inclusion of step-by-step patch recipes, and extensive web-based media content including tutorials, demonstrations, and additional background information. Sam McGuire and Nathan van der Rest guide you to master synthesis and transcend the technical aspects as a musician and artist.

Synths are presented using a multi-tiered system beginning with basic instructions for all common synth techniques. Historical information is included for each type of synth, which is designed to help you understand how each instrument relates to the bigger picture. Advanced level instruction focuses on modern implementations and on mobile devices, with special focus on performing and practical usage. The goal of *The Musical Art of Synthesis* is to bring all of the different types of synthesis together in the same discussion and encourage you to see the similarities and differences that force you to gain a better overall understanding of the synthesis process.

Key features of this title:

- This book will teach you how to put synthesizers to use with easy-to-use synth patch recipes.

- Using a unique, multitiered approach applicable to the level of equipment in use, this publication introduces concepts that apply to a wide range of hardware/software synthesizers.
- A robust companion website, featuring video demonstrations by synthesizer experts, further supports the book: www.focalpress.com/cw/mcguire.

Sam McGuire, an active audio engineer, composer, and University of Colorado Denver faculty member, is the primary author on three audio technology books and multiple audio software instructional videos. He has scored and mixed multiple award-winning documentary and feature films.

Nathan van der Rest is a University of Colorado Denver graduate student and a professional audio engineer and composer.

THE MUSICAL ART OF SYNTHESIS

Sam McGuire and
Nathan van der Rest

Focal Press
Taylor & Francis Group

NEW YORK AND LONDON

First published 2016
by Focal Press
70 Blanchard Road, Suite 402, Burlington, MA 01803

and by Focal Press
2 Park Square, Milton Park, Abingdon, Oxon OX14 4RN
Focal Press is an imprint of the Taylor & Francis Group, an informa business

Library of Congress Cataloging-in-Publication Data
McGuire, Sam, author.
   The musical art of synthesis / Sam McGuire and Nathan van der Rest.
      pages cm
   1. Synthesizer (Musical instrument)—Instruction and study.   2. Electronic music—
Instruction and study.   I. Rest, Nathan van der, author.   II. Title.
   MT724.M36 2016
   786.7'4—dc23
   2015006003

ISBN: 978-1-138-82978-7 (pbk)
ISBN: 978-1-138-82977-0 (hbk)
ISBN: 978-1-315-73759-1 (ebk)

Typeset in Palatino
By Apex CoVantage, LLC

CONTENTS

PREFACE

Most musical instruments are locked into sounding a certain way. In a blindfolded listening test, you could identify a piano from a guitar. Most of the time, you could also tell if an instrument is a synthesizer; but because of the wide-ranging sonic palette available in synthesis, you might not be able to guess correctly. Perhaps it sounds like a gentle breeze or like the crickets on a summer evening. It could sound like a robot from the future or a flute from medieval times. There are synths that sound like electric guitars or fog horns. Synths are instruments that don't necessarily require musical ability in the traditional sense, but when you explore further, you'll discover that they have a soul in the same way that a Stradivarius violin does; both require the touch of a musical artist to bring them to life at their fullest potential.

Recipe Creation

It's easy to focus on the technical aspects of synthesizers because so many of them require advanced knowledge when programming, because many vintage synths are aging, and because you have to have technical skills to keep them fully functional. New synths with unique features and layers of complexity are released frequently, with hundreds of different synths currently available in the marketplace. How do you know which ones to use and how do you get the most out of the ones you already own? What does it take to master synthesis and transcend the technical aspects as a musician and artist? That's why this book exists, because we asked those questions and we want to share what we found with you. We didn't, however, approach it haphazardly; we want

Figure 0.1 Synth recipe.

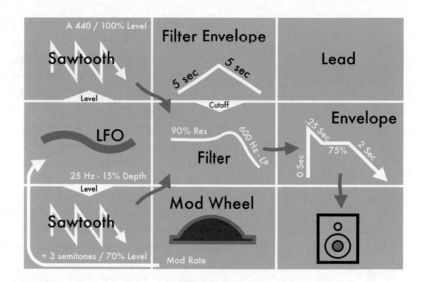

to be sure that you will transform as a person as you read this and that you return often to these pages as the learning process evolves.

There is a spectrum on which you fall when you use synthesizers and it starts with using presets and ends with the creation of patches from your own imagination. Unfortunately it isn't possible to put a lifetime of knowledge into a single book, but what we've done is something that is as close as we could make it. You see, Nathan and I have smelled the dust of original Arps and we have patched instruments crafted by Moog's own hands. We've tinkered for thousands of hours in the dim light with patch cables and knobs, and with each passing day, we have gained new insights into how each synthesizer lives and breathes. We remember the first synth we ever touched and catalog our lives in terms of the synths we've owned.

One day not too long ago, as we were preparing for the third annual Electronicatopia concert on the University of Colorado Denver's campus, we started designing a system of patch recipes that could be used to create sounds on a range of synthesizers. That's when we decided to make a master spreadsheet of every hardware synthesizer ever released, which includes a comparison of every major parameter. The idea is that with an analysis of the similarities of every synth, we can create a patch template which matches the largest pool of instruments and permits a single recipe to work universally. It might sound like a simple task with obvious results, but synthesizers are widely different and we had to overcome challenges that sometimes proved difficult.

A unified recipe system as a learning tool is going to change the way you understand basic synthesis. You will be able to create patches in Logic Pro that translate onto an Oberheim and into Ableton and beyond. It's not hard to learn the individual parameters and you can start tweaking your synths immediately, but our goal is for you to be proficient on the technical side of synthesis so that you can progress to explore the art and musicality of these amazing instruments.

The recipes are designed very carefully to provide insight into how things work together and to showcase patches for near instant use and also to teach you how to program your own. Recipes can be used in a variety of ways and much of this depends on your skill level. You can use them exactly as written to program sounds, or you can use them as launching pads for endless exploration. The recipes are designed primarily with subtractive synthesis as the primary focus, but in time, there is the possibility of expanding into some of the more difficult to document formats such as FM synthesis and Wavetable synthesis.

Synth Addiction

The past few years have experienced a synthesizer renaissance, and prices for some rare instruments have skyrocketed. Several instruments that were readily discarded a few years ago as new digital technology was released are now difficult to find and quite expensive to buy. A functional Moog Modular is impossible to come by and there are many others that new generations of synth users will only be able to read about online or in books. Some "vintage" synths which aren't as popular are available at discounted prices such as Yamaha's DX-7 or much of the Casio line.

There are also many new software and hardware synths for sale, such as the modular Eurorack synths, which are currently making their mark on the music industry. There are small synths and big synths, with software emulators and analog replications. Mobile phones and tablets have enabled ultracheap and portable synths that sound quite good and fit in your pocket. It feels like an epidemic that started with a small group of synthesizers and synth users, but it has spread to the general public across the world.

There is a bug that many synth owners catch which often turns into an unstoppable addiction and causes perfectly rational people to sell their possessions and obsessively stalk eBay for synthesizers. This addictive behavior fuels the fire for manufacturers to make new synths and keeps innovation alive in a field that could have easily folded when computers became strong enough to model classic synthesizers. One of the draws for hardware synths is that you have the tactile experience of turning knobs and patching cables.

Why are synths so popular? What are synths used for in the modern world? Some are used in the studio on records and others are used on stage for live performances. Some synths, however, are used for the sake of creative outlet by the synth purists that envision themselves as mad scientists hunkered over an invention brought to life after being patched together in the same way as Frankenstein's monster. These synthesizers are often controlled without musical keyboards, which are seen as an unnecessary crutch for the expert programmer; instead, the sounds are triggered by step sequencers, sample and hold modules, and other creative voltage control solutions.

How many synths do I have in my home studio? There are more than I care to admit and even if I did count them up, it would only take a few weeks before that number changes. My goal is to have one from every manufacturer and at least one of every type. It's getting close, but just like every junkie, I go through phases where I wonder if I should give them up. My newest rule is that if I don't use any one of them at least once a month then I should consider replacing it with something else. I am a mild case because there are musicians with entire rooms dedicated to their synthesizers and they have hundreds of synths and modules.

Hands-on Learning Process

If you read this book from cover to cover and expect to be a synthesizer expert, but you don't spend a lot of time experimenting and creating patches on an actual synthesizer, then you will be disappointed. We are going to teach you the names of all of the parts and we'll teach you how to create sounds: this is equivalent to memorizing vocabulary when learning a new language but never speaking it out loud. Speaking is an important part of learning a new language and is critical to mastering the necessary skills.

To get the most out of each section of this book, it is recommended that you follow along with a synth and put the concepts into immediate practice. This helps cement your understanding of the topics and then repetition reinforces that knowledge for future reference. You aren't going to be able to master your instruments in a single sitting, but if you take a consistent approach to using them and learning their quirks, then it's only a matter of time before you'll know them inside and out.

If you don't have a synthesizer that matches the type discussed in a particular chapter, then there are a couple options. Since it is critical that you have an instrument to practice with, you should either borrow or purchase an instrument. If you have a tablet or smart phone, then it is relatively cheap to buy app versions of each synth type, but if you are on a laptop or desktop, then one good option is to pick the right digital

Figure 0.2 Our favorite DAW, Logic Pro X.

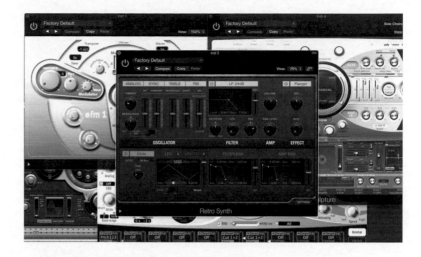

audio workstation (DAW) that has all of the included synths. In each chapter, we'll include information about which synths are recommended in the recipe areas, but a Google search will also help you find up-to-date information.

Synthesizers come in many shapes and sizes, and they use a variety of technologies in the synthesis process. The goal of synth recipes, and the unified system of notating them, is to bring all of the different types together in the same discussion; this encourages you to see the similarities and differences in such a way that forces you to walk away with a better overall understanding of the synthesis process.

SUBTRACTIVE SYNTHESIS

Subtractive synthesis has been used on thousands of records, and, in spite of its continued evolution, its essence remains the same at the very core. The reason it has continued to inspire musicians is because it can take electronic sounds and give them life through filters and envelopes that breath motion into the lifeless in order to create new, organic, exciting sounds or mimic the natural organic progression of acoustic instruments. Sounds are nothing less than magical when programmed by the experienced user.

The Synthesizer Database

When conducting research for this book, we thought it would be interesting to examine as many of the commercially available synthesizers that was physically possible in order to come up with averages of their capabilities. We compiled a comprehensive database that featured as many synths as possible (around 800 individual instruments) and outlined, in depth, their functionality. We examined things like the number of oscillators and low frequency oscillators, the number of filters and filter shapes, the number of envelope generators and their capabilities, the various wave shapes the synth could produce, the modulation routings, and the inclusion of various forms of performance control. We then averaged this database in relation to synthesis format so that we could provide you with the average number of synthesizers that feature a specific parameter throughout this chapter. As you will see in the coming pages, each parameter of subtractive synthesis will be complimented with the average percent of synthesizers that feature that particular function. So without further ado, let's begin with subtractive synthesis.

Subtractive synthesis starts with a sound. This sound is then filtered, modulated, mangled, and mashed, resulting in something beautiful. The starting sound comes from what is known as an oscillator, which is the heart of any synthesizer. Often underrated, the design of a synthesizer's oscillator ultimately determines the status that a synthesizer holds in history. Every parameter on a subtractive synthesizer—be it a filter, an envelope generator, a low frequency oscillator (LFO), or an amplifier—relies on the oscillator or oscillators in order to stand out amongst the vast and ever growing market of synthesizers. Take a MiniMoog for example; just listening to a single oscillator playing a triangle wave with no envelopes, resonance, or modulation is enough to send shivers down someone's spine. An average triangle wave is rather boring, but, for some reason, a MiniMoog's triangle wave connects with its users and instantly inspires them to reach out and turn every knob, opening up a world of creativity. The fact that every single waveform on a MiniMoog elicits this same response is the very reason that the MiniMoog became one of the most famous subtractive synthesizers in history.

> *Based on the information collected, subtractive synthesizers feature three oscillators on average.*

So what makes an oscillator so iconic? The functionality and features of an individual oscillator definitely play a role, but the sound of the individual waveforms an oscillator creates is where its iconic status ultimately rests. Let's examine what makes up a waveform as well as what makes a sawtooth wave sound different from a triangle or a square wave.

Harmonics

A waveform consists of a fundamental frequency and a series of additional frequencies called harmonics. The fundamental frequency is the frequency of the note being played. For example, the middle "A" note has a fundamental frequency of 440Hz, but most instruments do not produce just a single frequency. When a pianist strikes the middle "A" note on a piano for example, 440Hz is not the only frequency that is heard. Many higher frequencies, called overtones, emit from the piano as well. In a piano, these overtones are caused by the way the string vibrates, the resonance of the body of the piano, and other strings resonating in conjunction with the string being struck. The frequencies above 440Hz that emit from a piano when the middle "A" is struck are multiples of the fundamental frequency; these are called harmonics. The harmonics are produced in what is called the harmonic series, which is a pattern of harmonics that occurs naturally with musical instruments. In the harmonic series, the first harmonic is the fundamental frequency. Using our "A" 440 example, 440Hz is the first harmonic of the harmonic series. The second harmonic is twice the fundamental, or 880Hz. The third harmonic is three times the fundamental, or 1320Hz, etc.

Although the harmonic series is present in every instrument or sound, the amplitude at which these harmonics are heard vary from one instrument to another due to the material the

Figure 1.1 The harmonic series as demonstrated with vibrating strings.

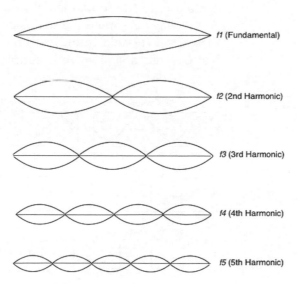

f1 (Fundamental)

f2 (2nd Harmonic)

f3 (3rd Harmonic)

f4 (4th Harmonic)

f5 (5th Harmonic)

instrument is made of, as well as the way in which sound is generated (strings, resonant tubes, membranes, etc.). These differences explain why instruments have unique sounds; a piano sounds different from a tuba which sounds different from an accordion and so on and so forth. This difference in tonal quality is called timbre. Synthesizers are such an attractive instrument because they can change timbres in an instant. The user is free to emulate a natural instrument or make a sound that is so earth shattering that it questions the very meaning of what is musical, all with the turning of a few knobs.

Waveforms

When dealing with synthesizers, each waveform has a unique sound due to the varying harmonic content that is produced. This is why a square wave sounds worlds apart from a sine wave. Let's take a look at the harmonic content of the most common waveforms found in subtractive synthesis in order to better understand the tonal quality of each of these waveforms.

Sine waves are the most simple of the waveforms. They contain only the fundamental frequency and no harmonics. A sine wave is something that is unique to synthesis because there are no nonelectronic instruments capable of such a tone. Just because a sine wave is simple does not mean that it is

Figure 1.2 The standard synthesis wave shapes.

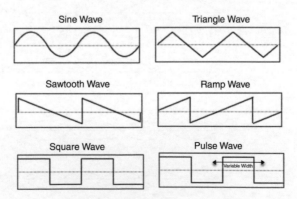

boring or less than useful. A sine wave, when placed in the audio spectrum correctly, will rumble the floors of a venue or pierce through the densest of mixes.

Take a Roland TR 808 drum machine for example. The 808 uses analog subtractive synthesis to create its drum sounds. The iconic kick drum sound of an 808 has destroyed sub-woofers and left an everlasting impression on anyone who has heard it thanks to its huge sine wave "ring" on the falling end of its sound. Sine waves have also long been coveted as a foundation for the largest of synth bass sounds. The extremely memorable bass sound on Nine Inch Nail's "Closer" would not have been possible without a sine wave in the mix.

Based on the information collected, 34% of subtractive synthesizers feature audio oscillator sine wave generation.

Triangle waves sound fairly similar to sine waves but with a bit more harshness. A triangle wave contains the fundamental frequency as well as all odd harmonics. This means that a triangle wave is made up of the first harmonic, or fundamental frequency, the third harmonic, the fifth harmonic and so on. The rate at which the higher harmonics drop in amplitude is proportionate to the inverse square of the harmonic number. For example, the third harmonic is 1/9th the amplitude of the fundamental and the fifth harmonic is 1/25th the amplitude of the fundamental. This rapid decline in the amplitude of the harmonics causes very few harmonics to be audible, which is why the triangle wave sounds similar to the sine wave. The few harmonics that are audible separate the triangle wave from the sine wave. As stated earlier, the MiniMoog's triangle wave is one of the many reasons the MiniMoog is still being talked about today. When overdriven into a filter, the triangle wave adds everything a sine wave is capable of adding to a sound, but also adds a new dimension of grit and glam that is only attainable through this means of synthesis.

Based on the information collected, 48% of subtractive synthesizers feature audio oscillator triangle wave generation.

Sawtooth waves have a very distinct, raspy sound quality. A sawtooth wave can scream and it pierces through any sound it's up against. At the same time, however, a sawtooth can be filtered back to create soft, delicate, and smooth sounds that can be placed in the most "moody" of R&B songs. Sawtooth waves get their name from their resemblance of a physical sawtooth when viewed on an oscilloscope. Unlike a triangle wave, a sawtooth wave contains the fundamental frequency as well as both even and odd harmonics. The rate at which the harmonics drop in amplitude is inversely proportionate to the fundamental rather than the inverse square of the fundamental like in a triangle wave. This means that the second harmonic is $1/2$ the amplitude of the fundamental, the third harmonic is $1/3$ the fundamental, the fourth harmonic is $1/4$ the fundamental and so on. Due to the slower decline in amplitude of the harmonics as well as having the even and odd harmonics audible, the sawtooth wave is extremely rich sounding and is very useful in subtractive synthesis. Due to the rich and harsh tonal quality of the sawtooth wave, it is extremely well suited for re-creating bowed string sounds such as cellos and violins and, as stated earlier, piercing lead and punchy bass synth sounds. The sawtooth wave is a staple amongst synthesizers and no instrument besides a synthesizer can come close to re-creating it.

Based on the information collected, 68% of subtractive synthesizers feature audio oscillator sawtooth generation.

Ramp waves, or reverse sawtooth waves, are simply backwards sawtooth waves. Rather than the wave peaking and then sharply sloping down, the ramp wave sharply slopes up and then peaks and drops back down to zero. The ramp wave contains the exact same harmonic content as the sawtooth and sounds identical. However, it is when using the oscillator as a control source for an LFO or as an envelope that ramp waves differentiate themselves from sawtooth waves and become one of the most overlooked and coveted of the waveforms. Mark Mothersbaugh of DEVO sent his MiniMoog back to the engineers at Moog in order to have it customized in order to produce a ramp wave. Mark's customized MiniMoog can be heard in the extremely memorable synth parts of DEVO's "Smart Patrol/Mr. DNA."

Based on the information collected, 41% of subtractive synthesizers feature audio oscillator ramp wave generation.

Next to sawtooth waves, square waves are the most recognizable of the waveforms. Square waves have a rich but hollow sound quality. Their name is derived from their square-like appearance on an oscilloscope. Like the triangle wave, square waves are made up of the fundamental frequency as well as all odd harmonics meaning they contain the first harmonic (fundamental), third harmonic, fifth harmonic, and so on. Unlike a triangle wave however, the rate at which the harmonics drop in amplitude are inversely proportionate to the fundamental. This means that the third harmonic is 1/3 the amplitude of the fundamental, the fifth harmonic is 1/5th the fundamental, the seventh is 1/7th the fundamental and so on. Because of the square wave's richer sonorities of harmonics, it imparts a full tonal quality to any synth sound that can only be described as legendary. Keith Emerson's synth solo towards the end of Emerson, Lake & Palmer's "Lucky Man" are perfect examples of what a square wave can add to a screaming lead line. The alternative band Passion Pit makes extensive use of the square wave on most of their synth parts throughout their entire catalog of music. The square wave is truly a synth staple and, for that reason, it can be found on almost 99% of synthesizers from the most basic and budget models up to the most expensive.

A pulse wave is, in essence, a variable square wave. Like a square wave, the pulse wave contains the fundamental and all odd harmonics with the same inversely proportionate harmonic amplitude drop. The difference is that the width of the wave in the positive and negative direction is variable. One of the most sonically pleasing and sought after sounds in synth history is created by modulating the width with an LFO. Typically, this variable width is continuously adjustable (from so narrow that it's not audible to all the way up to a full square wave), but some synthesizers have set width amounts. The song "Warp" by The Bloody Beetroots and Steve Aoki makes great use of pulse width modulation.

Based on the information collected, 85% of subtractive synthesizers feature square or pulse wave generation.

A Note on "Hyper" Waves

Many current manufacturers are including variations of the basic waveforms into their synthesizer designs and are labeling these new variations with the descriptor "hyper"; such as "hyper-saw" or "hyper-square" waves. These new waveform variations typically use wave folding or wave multiplying technology in order to alter the tonal characteristics of these waves to give the user more possibilities for sound creation.

Wave folding technology works by "folding" the wave back on itself causing more harmonics to be present at different levels. Wave folding can be thought of as the opposite of a filter; whereas a filter takes harmonics away from a sound, wave folding devices add harmonics into a sound. Wave multiplying technology multiplies the wave and usually shifts the pitch of the multiplied waves. These features not only give the user more sonic capabilities, but can make a single or dual oscillator synth sound like it has more oscillators, which is a great benefit to some of the low-cost models. The extremely popular analog subtractive synth from Arturia, called the MiniBrute, utilizes these hyper waves for all of the waveforms of the single oscillator, which makes it an extremely powerful single oscillator budget synthesizer.

Another hyperwave generation technique that warrants a discussion is wave shaping. Wave shaping produces similar results to wave folding, but rather than folding the wave back onto itself, wave shapers actually change the shape of simple waves, which creates sharper corners and increases harmonic content. The most common use of wave-shaping technology can be found in guitar distortion pedals and fuzz boxes. The clean guitar signal present at the device's input is amplified and forced to clip, which adds in newer harmonics, creating the distorted sound at the output. Wave-shaping technology is rarely seen outside the modular synthesizer world, but it is beginning to surface in some of the new analog synths due to the high demand for distorted synth sounds.

Different Types of Oscillators

Now that we have explored how oscillators and their waveforms make or break a synthesizer, let's examine the different types of oscillators that are available. When dealing with subtractive synthesis, there are three main types of oscillators one can expect to encounter—voltage controlled oscillators (VCO), digitally controlled oscillators (DCO), and software-based oscillators. These three types of oscillators are by no means the only type found on subtractive synthesizers, but they are by far the most common. Similar to the waveforms themselves, each of these types of oscillators imparts unique features and tonal qualities onto the synthesizer.

Voltage Controlled Oscillator (VCO)

The VCO is an oscillator that gets its pitch information from voltage input. Although many people experimented with different types of ways to control oscillators in the early years of synthesis, Bob Moog is most often credited with introducing the voltage controlled oscillator. Moog's early oscillators introduced the 1volt/octave standard found in most analog synthesizers both past and present. This standard divided a single volt into 12 parts to accommodate the 12-tone Western scale. This means that each half step is related to a 1/12-volt change. This 1volt/octave standard not only made it possible for synthesizers to produce tonal music, but also made it possible for synthesizers to communicate with each other. In the days before MIDI (and still today), analog synthesizers featured control voltage (C.V.) inputs and outputs. This means that a user could connect two or more synthesizers together and control each of them with a single keyboard.

Bernie Worrell used three separate MiniMoog synthesizers connected through C.V. to achieve the gigantic synth bass sound on the Parliament song "Flashlight." This type of connectivity would not have been possible without the 1volt/ octave standard.

It is important to note that not all synthesizers use the 1volt/ octave standard. Synths like the Korg MS-10 and MS-20, as well as Buchla synths, use their own Hz/octave standards. Connecting synthesizers together in order for them to communicate with one another has been, and still is, a must-have skill for the electronic musician. By having a multitude of synthesizers connected together, an artist can create an orchestra

Figure 1.3 Visual representation of the 1volt/octave standard.

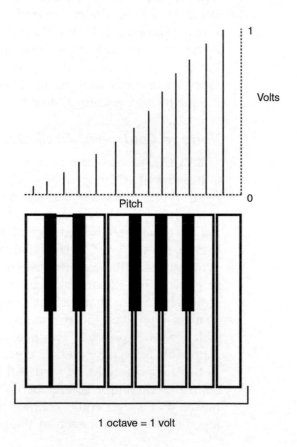

of synthesizers, all playing together in perfect rhythm and harmony, allowing one user to be the composer, conductor, and performer of his or her synthesized symphony.

Although VCOs are coveted for their "warm" tonal quality, they oftentimes drift in and out of tune and are susceptible to temperature and environmental changes, making them quite finicky to use. For this reason, the digitally controlled oscillator was introduced.

Digitally Controlled Oscillator (DCO)

DCOs get their name because they are controlled digitally, typically through integrated circuit chips (ICs). Although forms of DCOs were present in some of the early monophonic analog synthesizers, it wasn't until the introduction of polyphonic synthesizers like the Sequential Circuits Prophet 5 that DCOs really came into the spotlight. A polyphonic synthesizer is a synth that can play multiple notes at the same time. In hardware analog synthesizers, polyphony is achieved by having a separate synth voice for each note. Take the Prophet 5 for example, which has five voices of polyphony with each voice containing two oscillators, two envelope generators, a filter, and an amplifier.

At the time of the Prophet 5's creation, having ten separate VCOs all staying in stable tuning while tracking the keyboard perfectly would have been an engineering nightmare and so it made more sense to use DCOs, which are far more stable. DCOs were used extensively through the eighties and early nineties and are still found on new synthesizers today. In fact, Van Halen's iconic polyphonic synth sound on the song "Jump" would not have been possible had the Oberheim OBXa synthesizer not utilized digitally controlled oscillators.

Software-Based Oscillators

Software-based oscillators are typically found in software or VST (Virtual Studio Technology) synthesizers. There are a few hardware synthesizers such as the Arturia Origin and Access Virus line that also utilize software oscillators. As can be assumed, software-based oscillators use digital signal processing to produce tones. Most software oscillators are designed to emulate physical oscillators, but a few soft synths such as Logic's Sculpture utilize the capabilities of software in order to create new and interesting sounds through instrument modeling.

Using Oscillators

Now that we have covered the technical details of oscillators and the waveforms they produce, we can explore how oscillators are used in the sound creation process. When creating a sound, the user combines the onboard oscillators in the mixer section in order to create huge, rich walls of sound or delicate, beautiful sounds, which will then be filtered and modulated. Although the way in which the user utilizes the oscillators is solely up to the user, many parameters are provided on oscillators in order to aid in the sound creation process.

Tune/Detune

Sound creation starts with the pitch of the individual oscillators. Each oscillator can be set perfectly in tune with the others in order to create what effectively sounds like one mega oscillator, or the user can slightly detune the oscillators from each other resulting in a thick phasing, animated sound. Changing the octave of one of the oscillators (usually referred to as "range"—measured in footage in reference to organ pipe lengths) will result in a sound so large, it will be the dominate sound in any mix. The user can also tune one of the oscillators up or down an increment like a third or a fifth in order to create harmonies.

> *Based on the information collected, 74% of subtractive synthesizers feature oscillator detune capabilities.*

Oscillator Sync

By listening to the synth solo on Herbie Hancock's "Chameleon," one understands the sheer depth of what oscillator sync

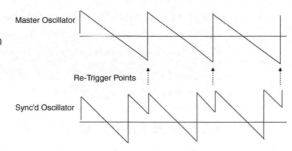

Figure 1.4 With oscillator sync engaged, the master oscillator retriggers the slave oscillator each time its wave cycle restarts.

can add to a synth patch. When oscillator sync is engaged, the first oscillator retriggers the second oscillator's wave when the first oscillator's waveform falls back to zero. This feature allows the user to adjust the second oscillator's pitch without getting the oscillators out of tune. The timbre of the sound will change depending on how the second oscillator is tuned, which creates extremely interesting sonic capabilities. Understanding and using oscillator sync is one of the things that separates the novice from the experienced. Oscillator sync is an essential tool in the synthesist's sonic toolbox.

Based on the information collected, 56% of subtractive synthesizers feature oscillator sync capabilities.

Sub-oscillator

Many synthesizers feature what is known as a sub-oscillator. A sub-oscillator clones the output of an oscillator and transposes it down one or two octaves. This allows the user to add an extra low end to a sound without using an entire oscillator to do so. A well-placed sub-oscillator thickens a synth patch to such a degree that once disengaged, it leaves users thinking their synths have become small and thin.

Based on the information collected, 28% of subtractive synthesizers feature sub-oscillators.

Noise

Noise is a crucial aspect of synthesis. Many synthesizers have an onboard noise circuit that produces white or pink noise.

13

The user can mix the noise with the oscillators when creating drum sounds, or in order to add a pseudo-distortion to the sound that is unique from any type of distortion pedal.

Based on the information collected, 76% of subtractive synthesizers feature noise generation.

Spectral Shaping (Filtering)

When thinking about a subtractive synthesizer, be it an analog monster such as an ARP 2600 or a modern DSP based beauty such as the Roland Aria System 1, it is impossible to imagine not reaching for the cutoff frequency knob of the filter and rolling it back to produce that ever-amazing filter sweep. This has become one of the most instantly recognizable sounds in music, and, for some reason, it seems to sound good in almost any musical setting. If the oscillator is a synthesizer's heart, then the filter is its diaphragm, lungs, and vocal chords—always breathing life into any patch, while giving the synthesizer a unique and distinguishable voice.

Bob Moog introduced the resonant low pass filter into his modular systems in the 1960s. Once he unveiled the 904a Transistor Ladder Filter, a paradigm shift took place, resulting in the landscape of synthesis that we know today. Every

Figure 1.5 The standard synthesis filter shapes.

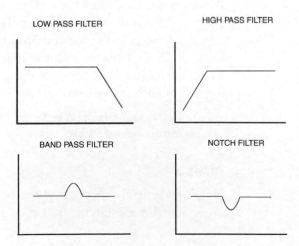

LOW PASS FILTER

HIGH PASS FILTER

BAND PASS FILTER

NOTCH FILTER

synthesizer company since Moog's unveiling has included a resonant low pass filter into their designs. The very fact that the Moog 904a filter was the only module for which Moog filed a patent shows the importance of the resonant low pass filter. Although the low pass filter is by far the most common filter type found in synthesizers, there are others and it is worth examining the differences of each one.

Low Pass Filter

As stated earlier, the low pass filter is the most common type of filter when dealing with subtractive synthesis. A low pass filter works by "rolling off" or cutting all frequencies above a user-specified point known as the cutoff frequency. The rate at which the filter reduces the amplitude of these frequencies is called its slope and is measured in dB/octave. The classic Moog filter is a 24dB/octave filter, which means all frequencies above the cutoff frequency are attenuated by 24dB at every octave. Another classic filter slope is a 12dB/octave filter. With a 12dB/octave filter, all frequencies above the cutoff frequency are attenuated by 12dB at every octave. Oftentimes, the filter's slope is labeled in "poles." One pole loosely relates to a 6dB/octave attenuation. Therefore, a 12dB/octave filter is known as a two-pole filter and a 24dB/octave filter is known as a four-pole filter. Some synthesizers have the option to switch the filter's slope in order to have more control over the end sound. Low pass filters are used to create the instantly recognizable filter sweep. Many synthesizers even allow external audio to be patched into the synthesizer's filter in order to perform this filter sweep on drums, vocals, or the song as a whole. Daft Punk uses low pass filters in order to filter all the instruments of a song back right before a huge climax in a large number of their songs. An example of this type of filtering can be found on the intro of Daft Punk's song "Around the World." The low pass filter is truly an icon in subtractive synthesis and will remain an icon for the foreseeable future and beyond.

> *Based on the information collected, 97% of subtractive synthesizers feature a low pass filter shape.*

High Pass Filter

The next most commonly found filter on a subtractive synthesizer is a high pass filter. A high pass filter (HPF) is the exact opposite of a low pass filter. In an HPF, all frequencies below the cutoff frequency are attenuated. HPFs are found alongside low pass filters on synthesizers and although not typically user selectable, high pass filters can have different slopes as well and follow the same dB/octave formula found on low pass filters. The Korg MS-20 synthesizer is famous for its filter section because it offers a resonant low pass alongside a resonant high pass filter. Extremely screeching leads can be produced when properly using a high pass filter. Although typically overlooked for bass patches, using a resonant high pass filter with just the slightest amount of filtering can produce one the largest synth bass sounds imaginable due to the resonant frequency being on the low end of the audio spectrum, but while still allowing all the bright, high-end sound to pass through. The Chemical Brothers are known to use the high pass filter of their MS-20s on most of their songs. An example of the screeching sounds only a high pass filter can provide can be found in the introduction of The Chemical Brother's "We are the Night" track. The high passed MS-20 is the screeching modulated sound that is heard while the rest of the song is low passed in the background.

Based on the information collected, 53% of subtractive synthesizers feature a high pass filter shape.

Band Pass Filter

Another filter found on some subtractive synthesizers is known as a band pass filter. A band pass filter (BPF) is a combination of a high pass filter and a low pass filter. The band pass filter attenuates all frequencies above and below a set "band" of frequencies centered on the cutoff frequency. A band pass filter often has a fixed bandwidth, meaning the amount of frequencies around the cutoff frequency that won't be attenuated is a fixed amount. Band pass filters are great for creating nasally thinner sounds such as oboes, saxophones, and other reed instruments.

Based on the information collected, 37% of subtractive synthesizers feature a band pass filter shape.

Band Reject or Notch Filter

The final filter type usually associated with subtractive synthesizers is known as a band reject or notch filter. A band reject filter is the exact opposite of a band pass filter in that the frequency band around the cutoff frequency is attenuated while leaving the frequencies on either side to remain unaffected. Although typically used in music production as a means to eliminate problem frequencies, the notch filter, when properly used on a synthesizer, can produce interesting sounds not otherwise attainable.

Based on the information collected, 24% of subtractive synthesizers feature a band reject or notch filter shape.

Using Filters

Like with anything on a synthesizer, the way in which the user utilizes the filter is completely up to the user. Most filters have predetermined parameters available to the user in order to aid in the sound-creation process. Let's take a look at some of the commonly found parameters of a subtractive synthesizer's filter in order to better understand what they do.

Cutoff Frequency

If you are only going to adjust one parameter on a synthesizer, chances are this is it. The cutoff frequency determines the frequency at which the filter will start working and it is this control that produces those infamous filter sweeps. The cutoff control is usually a prominent knob or slider in the filter section with easy access so that the user can manually adjust it in real time.

Figure 1.6 Cutoff frequency.

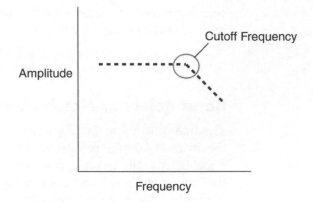

Resonance

Ever since Moog released the 904a transistor ladder filter in the 1960s, resonance, sometimes known as "peak" or "emphasis," has been a nearly universal parameter on all synthesizer filters. Resonance is a feedback circuit that feeds the cutoff frequency back into the filter, causing that frequency to jump above the center line and boosting the cutoff frequency and the adjacent frequencies. As the user turns up the resonance knob, the filter begins to ring or "resonate" at the cutoff frequency. Some filters self-oscillate when turned up, producing a sine wave at the pitch of the cutoff frequency. When used in conjunction with the filter cutoff, resonance imparts an amazing, animated sound onto any synth patch. Resonance can be thought of as the secret ingredient that makes filters so amazing and memorable.

> *Based on the information collected, 94% of subtractive synthesizers feature resonant filters.*

Filter Voltage Control

Like with analog oscillators, analog filters are controlled via control voltages. These voltages are used to set the cutoff frequency without the need for the user to manually adjust the cutoff frequency knob. By patching an LFO or other modulation source into a filter's control voltage input, the user can be

Figure 1.7 Resonance creates a "bump" in the frequencies around the cutoff frequency.

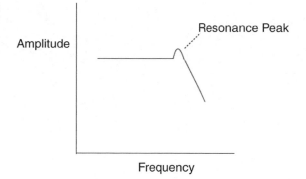

free to play more complex lines while the filter automatically sweeps.

Envelope Amount

The envelope amount control determines the depth at which an envelope generator affects the filter's cutoff frequency. Although the way in which the envelope controls the filter is covered in depth later in the chapter, it is important to know that most filters allow the user to determine just how much an envelope controls the filter. Most filters even allow the user to decide the polarity of how the envelope controls the filter, meaning that the envelope can control the filter in a positive or negative manner. By using an envelope to control a filter, the user can set the exact way the filter opens and closes every time a key is struck. Although some lower-end analog synthesizers do not allow envelope control over the filter, it has become an almost universal feature on synthesizers today.

> *Based on the information collected, 88% of subtractive synthesizers feature an envelope amount control over the synthesizer's filter.*

Key Follow

Because analog filters are controlled via control voltage, the same controllers used to determine pitch on an oscillator can be used to determine the cutoff frequency of a filter.

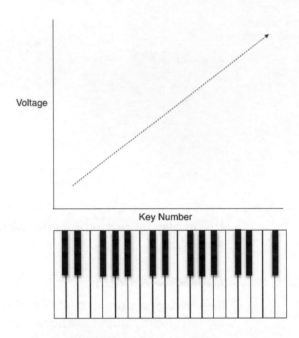

Figure 1.8 Visual representation of key follow.

This means that the synthesizer's keyboard that produces higher voltages the higher up one plays can control the filter, and so most filters offer a function called "key follow" or "keyboard control." When engaged, this function increases the filter's cutoff the higher one plays on the keyboard, causing higher notes to be brighter and lower notes to be darker. Since this is such a sought-after feature on analog synthesizers, most modern digital and software synthesizers offer this feature using digital control instead of voltage control. Key follow can be used in more extreme degrees as a way to have a bass-sounding filter response at the lower end of the keyboard while having a lead-sounding filter response at the higher end. When used more subtly, key follow will impart an ever-evolving filter response as one plays across a keyboard. One lesser-known way in which to use key follow is to crank the resonance to its highest degree, causing the filter to self-resonate; then, by tuning the cutoff frequency so the filter resonates at a specific note, the user can "play the filter," meaning that the resulting sine wave will track up and down the keyboard and allow the user to have a fully functional sine wave generator.

Based on the information collected, 77% of subtractive synthesizers feature key follow functionality.

Filter Selection

Rather than having individual filters for low pass, high pass, band pass, and band reject filter types, many synthesizers allow the user to switch filter types using the onboard filter, allowing the synthesizer to have one, all-encompassing filter. The Oberheim multimode filter found on their analog synths is usually credited as being the first successful, multimode filter. That being said, it is not uncommon to see a high pass filter in conjunction to the main multimode filter on many synthesizers.

Amplitude Shaping (Amplifiers and Envelope Generators)

In an analog oscillator circuit, sound is always present at the output stage whether a key is pressed or not. Because oscillators continuously produce a pitch, it is necessary to have a circuit that only lets sound out of the output when a key is pressed. The circuit responsible for releasing sound to the output or withholding sound when the user specifies is known as the amplifier. In an analog synthesizer, the amplifier is controlled via control voltage in the same vein as the filter and oscillators. The amplifier allows sound to pass when a positive voltage is present and refuses sound from passing when no voltage is present. This voltage is known as a gate signal and is typically made up of a +5v pulse wave. If synthesizers just produced sound the instance a key was depressed and immediately fell silent the instant the key was released, they would neither be very musical nor beneficial to most users. Therefore, it is necessary to be able to control the way in which the synthesizer's sound begins, maintains, and falls. This type of sound control is achieved with what is known as an envelope generator.

Envelope Generators

When a pianist strikes a key on a piano, the sound does not maintain its full amplitude and energy throughout the duration of the key being depressed. Once a key is struck on a piano, the sound level quickly peaks and then falls and continues to fall slowly until it is inaudible. This type of sound characteristic is known in the synthesizer world as the instrument's envelope. An envelope generator replaces a gate signal with a modified gate signal in which the user specifies the rate at which the sound peaks and falls.

The most common type of envelope generator found on synthesizers is known as an ADSR envelope. ADSR stands for attack, decay, sustain, and release. The user specifies the length of each of these parameters in order to create a desired envelope for the sound. Although many synthesizers offer full ADSR envelope generators, some offer variations such as an AR envelope, which would just have attack and release controls, or an ADR envelope which would just have attack, decay, and release controls. This is especially true when synthesizers feature more than one envelope generator, in which case, one will usually be an ADSR while the others might be

Figure 1.9 The standard ADSR envelope generator shape and its parameters.

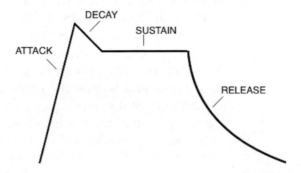

a variation. Let's take a look at each of these parameters in depth to better understand how they affect the overall sound of the synthesizer.

Based on the information collected, subtractive synthesizers feature an average of three envelope generators.

Attack

The attack portion of an envelope generator controls the speed at which the sound is heard when a key is struck. A short attack yields a sound that is heard at the instant the key is depressed, while a slow attack yields a sound that slowly rises in amplitude from when the key is depressed. The rate at which the attack of an envelope can be adjusted is different from synthesizer to synthesizer, but is usually adjustable from a few milliseconds to a few seconds. By simply slowing down the attack, beautiful string sounds can be created that rise and build to their climax slowly. Slowing down the attack time is also great for making long, warm synth pad type patches.

Based on the information collected, 96% of subtractive synthesizers feature an envelope attack control.

Decay

The decay portion of an envelope generator specifies the amount of time it takes the sound to fall from its highest peak to its sustain point. Like the attack parameter, the decay control can typically be adjusted from a few milliseconds up to a few seconds. Decay can be thought of as the amount of time it takes the piano to fall to its sustaining sound once the transient sound of the string being struck has died down. Setting a long decay time will help create an ever-moving and evolving synth pad sound.

Based on the information collected, 93% of subtractive synthesizers feature an envelope decay control.

Sustain

The sustain parameter of an envelope generator determines the amplitude at which the sound remains while the key is depressed after its initial attack and decay times. Unlike the attack and decay parameters, the sustain parameter is not measured in time. Instead, a lower sustain results in a signal with a lower amplitude once it reaches its sustain point, while a greater sustain results in a sound with a greater amplitude. Many budget synthesizers unfortunately neglect to include a user-adjustable sustain parameter on their envelope generators, even though the sustain parameter is extremely important in sound creation. When setting an extremely low sustain with a short attack and decay, a synth can produce extremely short percussive sounds great for drum creation, chirping rhythms, or complex sequences. When cranking the sustain to its highest level, the amplitude will stay at its highest peak for the duration of the key being depressed, resulting in huge walls of sound sure to rock the foundation of any piece of music the sound is inserted into.

> *Based on the information collected, 91% of subtractive synthesizers feature an envelope sustain control.*

Release

The final parameter found in an ADSR envelope is known as release. The release determines the amount of time it takes the sound to fall down to an inaudible level once the key is released. This is like the way that a piano still "rings out" even once the player releases a depressed key. Release time is another important parameter that is sometimes left out of budget synthesizers. The release parameter, when set to longer lengths, allows a user to have notes continue to sound until the next note is triggered, effectively mimicking the way a sustain pedal works on a piano when the pianist releases the pedal at the next note. When using the release parameter in this way, the user is free to move his or her hands up or down the keyboard in anticipation of the next phrase without the synthesizer abruptly going quiet when a key is released.

Setting a long release not only aids in creating long, moving synth pads, but also helps create screaming synth solos.

Based on the information collected, 87% of subtractive synthesizers feature an envelope release control.

Envelope Control over Filters

An envelope generator is used for more than just controlling amplifiers. One common parameter that is controllable via an envelope is the filter. In fact, many synthesizers contain a second ADSR envelope generator designed specifically to control the filter. Having an envelope available to control the filter allows the user to program filter sweeps and evolving filter sounds without having to manually adjust the cutoff frequency. Although the envelope controls are the same when controlling a filter, the results each parameter has on the filter are different than with an amplifier, so it is important to examine the effect each parameter bestows when controlling a filter.

Filter Envelope Attack

When using an envelope generator to control a filter, the attack determines the amount of time it takes the filter to reach the cutoff frequency once a key is pressed. By adjusting the attack time, the user effectively creates a filter sweep up to the cutoff frequency when striking a key. With a bit of resonance and a slightly longer attack time, the filter opens shortly after the note sounds, creating an almost wet, bouncy type of sound that lends itself to bass patches beautifully. This type of sound is heard on Prince's "Controversy" song. A long attack time slowly increases the cutoff frequency, allowing for extremely animated synth pad and lead sounds. Adjusting the attack time of a filter also aids in creating brassy sounds.

Filter Envelope Decay

When controlling a filter, the decay control determines the amount of time it takes the filter cutoff frequency to fall from its highest level right after the attack, to its set sustain point. This parameter allows the user to have a filter that sweeps down after a key is pressed; this allows it to pierce through a mix at the instant the note is heard and then fall back into the background while still being audible. Like with an amplifier envelope, the decay works in conjunction with the sustain parameter.

Filter Envelope Sustain

The sustain parameter in regards to the filter envelope determines a new cutoff frequency that the filter will remain at while a key is being depressed after the initial attack and decay times have run their course. By setting a low sustain with a long decay, the filter slowly closes while a key is being held, creating beautiful, evolving pads and lead sounds. A low sustain with a short decay yields extremely sharp and percussive sounds that, when mixed with resonance, are used to create slap bass or kick drum-like sounds.

Filter Envelope Release

The release control of a filter envelope determines the speed at which the filter closes once a key is released. Adjusting each parameter of a filter envelope generator allows the user to create ever-evolving filter responses that can sound either beautiful and musical or extremely crazy and exciting. By having an ADSR envelope for both the filter and the amplifier, the possibilities are endless on the sounds that can be created.

Envelope Control over Pitch

One final use for an envelope generator found on certain synthesizers is the ability to control pitch. A pitch envelope is used for creating sound effect-type sounds but it can also be quite musical. When using a pitch envelope, the attack control determines the speed at which the oscillator rises in pitch until it reaches the set pitch of the oscillator. The decay determines the speed at which the pitch will fall from the highest point to the user-set sustain pitch. The sustain control determines the frequency at which the oscillator will stay while a key is being depressed. And finally, the release control determines the time it takes the pitch to fall from the sustain pitch down to an inaudible range once a key is released. The ability to have so much control over every aspect of the sound is one of the main reasons synthesizers are able to create the sounds that they can.

A Note on Envelope Generators

It is important to understand that an envelope generator is the same exact circuit whether it's controlling an amplifier, a filter, or an oscillator's pitch—the only difference being the destination it is routed to. For example, when using a modular synthesizer, which uses patch cables to connect individual modules such as oscillators and filters together, the envelope generators one would use to control amps, filters, and oscillators would all be the exact same module. The number of destinations to which one could route an envelope in a modular system would be reliant on how many individual envelope generators were in the system. On a hardware, nonmodular synthesizer such as the MiniMoog, the synthesizer designer determines a fixed number of envelopes and destinations. In the case of the MiniMoog, there are two individual ADSR envelope generators with one permanently routed to the filter and one to the amplifier.

Modulation (LFOs and Sample and Hold)

An LFO, or low frequency oscillator, is an oscillator which produces frequencies below the audible range in order to control other aspects of the synthesizer. LFOs are the main source of modulation in a synthesizer, and like an envelope generator, the LFO can change the sound of various parameters on a synthesizer. In some synthesizers, LFOs can be used to trigger envelope generators in order to create precise repeating rhythms. LFOs are one of the most versatile features on a synthesizer because they can be used to add slight vibrato or tremolo onto a sound, or be used to create rhythmic mangling of a sound heard on the most intense of electronic music songs. The stereotypical "Dub-Step" wobble bass sounds are created using LFOs. Let's examine just what an LFO is and what it does in order to better understand how it can be used.

LFOs vs. Oscillators

As stated above, an LFO is an oscillator that produces frequencies that are lower than the audible frequency range. It should be understood that an LFO circuit is virtually identical to an oscillator circuit, with the only difference being that the frequencies they produce are lower. In the MiniMoog for example, there is no stand-alone LFO; instead, the third oscillator can be switched into "LO" mode, which lowers the frequency range of oscillator three in order for it to be used as an LFO. Because LFOs are identical to oscillators, all the wave forms of a traditional oscillator can be utilized in an LFO. In order to keep costs down, many synthesizer companies limit the amount of waveforms their LFOs can produce. That being said, higher-end synthesizers usually have LFOs that can produce most if not all waveforms that their oscillators produce. Therefore let's examine each of the standard synthesizer waveforms again, this time concentrating on what effect they will produce when used as an LFO.

Based on the information collected, subtractive synthesizers feature two LFOs on average.

LFO Sine Wave

A sine wave has gently rounded peaks and troughs, which correspond to the gentle rise and fall of the parameter being modulated at the speed of the LFO. The sine wave is often used to create vibrato and tremolo effects and is one of the most common LFO waveforms offered on both expensive and budget synthesizers. When used at their extreme, sine wave LFOs impart an intense wobble onto whatever parameter is being modulated.

Based on the information collected, 44% of subtractive synthesizers feature LFO sine wave generation.

LFO Triangle Wave

Next to the sine wave, the triangle wave is the second most common waveform found on LFOs. Similar to the sine wave, the triangle wave is used heavily for both vibrato and tremolo effects. Based off of the triangular wave shape, a triangle wave LFO imparts a sharper rise and fall onto the parameter being modulated, allowing for sharper vibrato in their more modest settings and harsher wobbles in their more extreme settings.

Based on the information collected, 80% of subtractive synthesizers feature LFO triangle wave generation.

LFO Sawtooth Wave

When used as a modulation source, the sawtooth wave imparts an immediate rise and then elongated fall onto the parameter being modulated. Sawtooth modulation sounds like rhythmic knife slices are being taken out of whatever they are modulating, and so sawtooth wave modulation is extremely effective when creating sound effects on a synthesizer.

Based on the information collected, 54% of subtractive synthesizers feature LFO sawtooth wave generation.

LFO Ramp Wave

The ramp wave is a reversed sawtooth wave and when used in oscillators, ramp and sawtooth waves are virtually indistinguishable. When used as a modulation source, however, ramp and sawtooth waves are very different. A ramp wave modulation imparts a rise and then instantaneous fall onto the parameter being modulated. Like sawtooth modulation, ramp wave modulation is extremely useful when creating sound effects on a synthesizer and effectively does whatever sawtooth modulation does, only reversed.

Based on the information collected, 40% of subtractive synthesizers feature LFO ramp wave generation.

LFO Pulse/Square Wave

Square wave modulation is often used to trigger envelope generators in a synthesizer. When used as a trigger source, a square wave LFO triggers the synthesizer to sound every time the wave cycle repeats. This means that instead of hitting a key on the keyboard repeatedly on every beat, the user can use an LFO to achieve this sound with exact timing and precision. Pulse/square wave modulation is not limited to envelope triggering, and a pulse/square wave can also be used to modulate any parameter that is available to be modulated. When used as a modulation source rather than a trigger source, a square/pulse wave imparts up/down modulation onto the parameter being modulated, producing anything from "glitchy," stabbing sound effects to rhythmic, pulsating sounds.

Based on the information collected, 78% of subtractive synthesizers feature LFO pulse or square wave generation.

LFO Pitch Modulation

A common destination that can be modulated via an LFO is pitch. Pitch modulation works by routing the LFO output into the oscillator's pitch control. When using pitch modulation, the pitch rises with the rising LFO waveform and falls with the falling LFO waveform. The amount that the LFO affects pitch can be determined with an LFO depth control, such as a knob or controller wheel. Pitch modulation creates a vibrato effect at lower depths and a harsh, frequency-modulated effect at higher depths. If oscillator sync is engaged and the LFO is set to modulate only the sync'd oscillator's pitch, some of the most exciting sounds in synthesis can be created.

LFO Amplifier Modulation

When used to modulate the synthesizer's amplifier, the LFO can be used to create a tremolo effect. When modulating an amplifier, the amplitude increases with the rise of the LFO's waveform and decreases with the fall of the LFO's waveform. At greater depths, the sound cuts in and out at the rate of the LFO. When the LFO's frequency is increased to near audio rates, the synth gets an almost distorted quality to it that sounds truly amazing.

LFO Filter Modulation

Another common LFO destination is the filter. When engaged, the filter opens with the rise of the LFO's wave and closes with the fall of the LFO's waveform. This means that an LFO can be used to increase and decrease the cutoff frequency of the filter, creating pleasing, long filter sweeps or "Wah-Wah" type effects. When filter resonance is increased, filter modulation imparts an extremely pleasing animated sound to the filter at lower depths and extreme glitchy sounds at more extreme depths.

LFO Pulse Width Modulation

Pulse width modulation (PWM) is a staple in subtractive synthesis. To achieve pulse width modulation the LFO must be routed to the pulse width control of one or more of the oscillators. The width of the pulse wave changes with the rise and fall of the LFO's waveform. Most higher-end synthesizers as well as budget synthesizers offer pulse width modulation due to its sonic qualities and high esteem held by users.

> *Based on the information collected, 74% of subtractive synthesizers feature pulse width modulation.*

Sample and Hold

Although not exactly an LFO waveform, many LFOs feature sample and hold functionality. A sample and hold circuit

Figure 1.10 Visual representation of sample and hold.

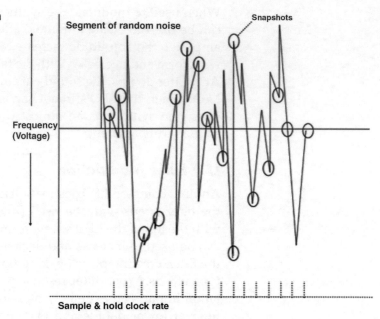

Segment of random noise

Snapshots

Frequency (Voltage)

Sample & hold clock rate

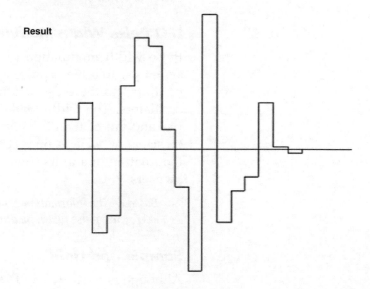

Result

takes voltage snap shots of a signal at its input stage and then applies them to a parameter of the synthesizer at a set rate. The most common signal typically fed into an LFO is white noise. Because white noise contains all frequencies, a sample and hold circuit takes a snapshot of a different frequency at every interval, resulting in a random sequence at the output stage. One way in which sample and hold is utilized is by feeding the output stage to an oscillator, causing random pitches to be produced at the rate of the sample and hold circuit. Another typical use of sample and hold is for the output stage to be routed to the filter, causing random cutoff frequencies to be produced. Arp was famous for its versatile sample and hold circuits on their synthesizers. When the late George Duke was Frank Zappa's keyboardist, he used the sample and hold of his Arp Odyssey on many iconic Zappa songs. Pete Townshend used the sample and hold circuit in conjunction with a sequencer patched into the filter on his Arp 2500 to create the iconic opening sound on The Who song "Baba O'Riley."

> *Based on the information collected, 58% of subtractive synthesizers feature sample and hold capabilities.*

Controlling a Synthesizer

In the early days of electronic music synthesizers, controlling pitch, especially the fixed pitches of a 12-tone Western scale, was troublesome. Many people were experimenting with different ways to produce set pitches. By far the most common type of synthesizer control is the keyboard (similar in look and function to a piano keyboard). Bob Moog is most often credited with introducing the traditional black and white keyboard into synthesis. At the same time that Moog was using keyboards to control his modular systems however, Don Buchla was experimenting with new and interesting ways to control his modular synthesizers in an effort to release the electronic musician from feeling trapped to tonal, Western music. The types of controllers that Buchla was designing included sequencers, push buttons, and metal touch plates. Although the right type of controller to use in order to control a synthesizer is solely up to the user and his or her wishes, the traditional keyboard is by far the most common.

Keyboards

Analog synthesizers are controlled via voltages known as C.V., and so analog synthesizer keyboards produce control voltages whether they are stand-alone keyboard controllers for modular synthesizers or built-in keyboards, such as is found on the Arp Odyssey or the Korg MS-20. Each key on an analog synthesizer typically produces two voltages, a 1/12-volt increment (or different volt/Hz/octave standard) for controlling pitch, and a +5v pulse for producing gates. On a stand-alone synth, such as the MiniMoog, the volt/octave voltage is internally routed to the oscillator for pitch, for filter, and for key follow, while the pulse voltage is internally routed to the envelope generators and amplifier. In many modern synthesizers, even modern analog synthesizers (such as the Arturia MiniBrute), the keyboard is connected to the oscillators and filters via MIDI. The keyboard is the most prominent pitch controller for synthesizers due to the ease in which users can play the synthesizer musically.

Figure 1.11 Traditional synthesizer keyboard as seen on the Sequential Circuits Pro-One synthesizer. Photo courtesy of www.switchedonaustin.com.

Sequencers

A sequencer is a device that produces note and gate information in order to control a synthesizer without the need for a user to physically play a keyboard. The most common type of sequencer is what is known as a step sequencer. An analog step sequencer usually has either 8, 16, or 24 steps that can each be programmed individually to produce a note in a given range (around three octaves). A step sequencer receives its clocking information from a gate pulse, either internally or from an external source. Step sequencers have the ability to place a mute on any of the steps to create complex patterns. Analog sequencers, both vintage and modern, produce control voltages, while modern digital sequencers typically produce MIDI notes. Because sequencers produce either control voltage or MIDI information, it is easy to integrate them into any synthesizer setup. Although early sequencers were stand-alone units or modules, many synthesizer companies began including them in their synths in the mid-1970s and still do today. Giorgio Moroder used sequencers extensively in his music, which in turn changed the course of electronic music and practically paved the way for modern dance music. Moroder's ground-breaking

Figure 1.12 Division 6 SQ18 Eurorack step sequencer.

sequences can be heard extensively in Donna Summer's song "I Feel Love."

Based on the information collected, 29% of subtractive synthesizers feature sequencers.

Arpeggiators

Like a step sequencer, an arpeggiator produces note and gate information in a step-like pattern. The main difference being that an arpeggiator gets its note information from keys being held down on a keyboard and reproduces these notes in a low-note-to-high-note order. Therefore, arpeggiators are far more limited in their ability to be programmed and are generally used as a means of live performance control rather than a stand-alone synthesizer controller. Arpeggiators get their name because they physically arpeggiate chords that are being held down on the synthesizer's keyboard.

Based on the information collected, 28% of subtractive synthesizers feature arpeggiators.

Other Means of Synthesizer Control

As stated earlier, designers like Don Buchla were experimenting with various types of synthesizer controllers at the time

Figure 1.13 Visual representation of an arpeggiator.

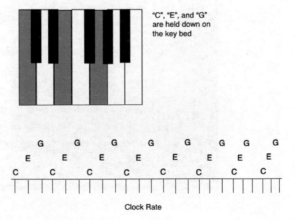

of the traditional keyboard's introduction to synthesis. These more esoteric controllers are very much still alive today and are available to artists who want to veer away from traditional, Western, tonal music. Modern Buchla synthesizers, for example, use touch plates that react to a variety of different factors, such as where a finger physically is on the plate, the amount of surface area a finger takes up, as well as pressure. Other controllers such as the Haken Continuum use ribbon technology, which allows a user to smoothly glide between notes by sliding his or her finger across the controller. Some synthesizers utilize a combination between esoteric control and traditional keyboard control. The EMS Synthi AKS, for example, used a touch plate printed to look like a traditional keyboard that responds to the 40Hz hum present in the human finger. The EDP Wasp synth used a similarly flat touch plate, but responds to heat rather than the 40Hz hum in the human finger. The amount of esoteric controllers on the market is vast and it would fill many chapters to detail each and every one.

Performance Control

Most synthesizers offer various means of performance control. These controls are designed in order to give the user maximum control over the synthesizer in order to perform live or in the studio. Let's examine some of the more common types of performance control on subtractive synthesizers.

Glide

Glide, sometimes referred to as portamento or glissando, allows the user to create a smooth slide between notes when playing on the keyboard. With glide engaged, the oscillators will rise in pitch, at a rate determined by the user, from the first note to the next when playing up the keyboard, and will fall in pitch from one note to another when playing down the keyboard. The glide function is similar to bending a guitar string from one note to another, meaning that one note slides in pitch to the next rather than just producing the first note and then producing the second. The glide time can usually be set between a few milliseconds up to a few seconds. Using the glide function while performing on a synthesizer imparts a musical attribute to the sound that, when applied correctly, can transform a synth line from being boring and stagnant to being exciting and moving. In more extreme cases, the user can set the glide time to extremely long lengths in order to create a sound that keeps slowly rising or falling in pitch as an effect.

> *Based on the information collected, 79% of subtractive synthesizers feature glide capabilities.*

Pitch Wheel

Most synthesizers feature a way in which to "bend" the pitch of an oscillator up or down. Most often, this control is seen in

Figure 1.14 MiniMoog Voyager pitch wheel (left) and modulation wheel (right).

the form of a wheel, but joysticks, levers, and knobs are not uncommon as well. Many synthesizer pitch wheels, joysticks, and levers are spring loaded; in this way, the user can quickly bend the pitch and let go of the controller so that the pitch will quickly return back to its starting pitch.

Based on the information collected, 60% of subtractive synthesizers feature pitch wheels.

Modulation Wheel

When using an LFO to modulate a certain parameter of a synthesizer, the user can set the depth of the modulation. Most synthesizers offer a wheel, joystick, or lever similar to the pitch control for setting the modulation depth. Unlike the pitch control, however, a modulation wheel is typically not spring loaded, so the user can increase the depth and leave it for as long as desired.

Based on the information collected, 52% of subtractive synthesizers feature modulation wheels.

Ribbon Controllers

Although not nearly as common as wheels, joysticks, and levers, ribbon controllers are found on certain synthesizers. A ribbon controller works by sliding a finger across its flat surface. A ribbon controller can usually be set to control a variety of parameters on a synthesizer, but pitch, filter, and modulation depth are the most common destinations for the ribbon controller to affect.

It is because of this vast amount of user-controllable features that subtractive synthesis has survived throughout the decades since its introduction. By learning what each parameter does and how it affects the sound, the user creates a physical connection between them and the instrument, which imparts a sentimental quality to the music that is produced. People have long coveted their synthesizers and held them in the highest regards like a classically trained violinist cherishes their violin. Music would not be where it is today had the electronic music synthesizer not come onto the scene and produced a paradigm shift in which the entire music landscape was altered.

Recipes

The following ten recipes have been handcrafted by us and are designed to help teach some of the theories and techniques mentioned in this chapter. For the beginner, these patches will provide you not only with an eclectic set of patches, but will also help instill some fundamental synthesis knowledge by physically re-creating these patches. For the experienced, these patches can be a jumping-off point as well as a source of inspiration for creating your own patches. Either way, these patches are designed to sound good, represent and feature various functions of a subtractive synthesizer, and provide you with some fun synth explorations.

Recipe 1: Punchy Bass

This first recipe is designed as a synth-heavy bass patch. The patch features a quick filter envelope, which causes the filter to close rapidly and only allowing for a small fragment of full, harmonic content. With added resonance, this filter envelope setting really makes the sound become animated. This patch is well suited for really any type of genre that requires a super synth bass.

Figure 1.15 Recipe 1—Punchy Bass.

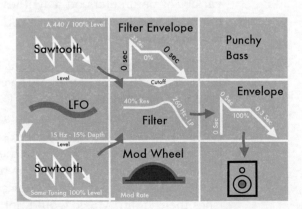

When re-creating this patch, the two oscillators will be set to sawtooth and square waves respectively. The two oscillators should be set perfectly in tune with one another. If desired, the two oscillators can be slightly off tune with one another by a few cents to thicken up the sound. This patch was designed as a bass patch so the oscillators should be set to 32' or 16' ranges. The low pass filter's cutoff should be set at 260 Hz in order to roll off much of the high-frequency content of the sound. Resonance should be set to around 40%, but this can be tuned to taste depending on your particular filter. This patch requires two separate envelope generators with one routed to control the amplifier and one routed to control the filter. Starting with the filter envelope, the attack control should be set to zero seconds, or as short as it can go. The decay parameter should be quite short, around 0.25 seconds. The sustain should be all the way down at 0% and the release should also be turned all the way down. By having such a short decay time with no sustain, the filter will be open right as a key is pressed, and then it will quickly close, resulting in a quick filter sweep that plays on the resonance setting with each key press. Moving onto the amplifier envelope, the attack and decay parameters should be set to zero seconds, while the sustain should be set to 100%. Finally, the release should be set to around 0.3 seconds. By having the sustain set to 100%, the amplifier will stay at its max amplitude for the duration of a key being depressed. The release is set slightly longer in order to ensure no pops will be heard when a key is released. This patch was designed without any glide, but a short glide time could be added to add a bit of an extra flair. An LFO routed to the oscillators pitch and controlled via a mod wheel should be set to about 15Hz with around 15% depth to add vibrato when desired.

If a bit more punch is desired, a bit of white or pink noise could be added at a low level. Because of the shape of the filter envelope, most of the high frequency of any added noise will be cut out quickly, resulting in a nice, low-end oomph at each key press. Additionally, a third oscillator producing

41

a sine or triangle wave could be added at an octave higher than the first two oscillators in order to add some extra beef to the sound.

Recipe 2: Evolving Lead

The second patch in our list is a slightly different take on a traditional lead patch. The filter will slowly open once a key is pressed, revealing more high frequency content as the sound progresses while a heavy dose of resonance will animate the sound and help it pierce through any mix. This patch was designed to fit greatly in an ambient or alternative setting, but will bode well in any R&B or even EDM track. This patch is extremely versatile; by simply adjusting the filter envelope's attack, the sound can take on whole new life.

When re-creating this patch, the two oscillators will be set to produce sawtooth waves. One of the oscillators will be tuned an octave below the other to create a rich tone. The low pass filter cutoff is set relatively low at around 236Hz, while the resonance is set extremely high at around 85%. The filter envelope's attack is set quite long at 2.5 seconds, while its decay is set longer—still at around three seconds. Sustain is set at 35%, while the release is set all the way up. This type of envelope shape will result in a filter that slowly opens once a key is pressed, and then slowly dies down to a more subdued

Figure 1.16 Recipe 2—Evolving Lead.

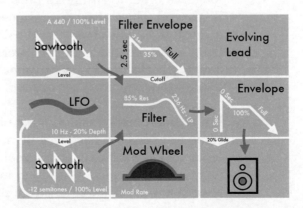

state until it slowly closes once a key is released. This type of filter envelope shape lends itself well to a lead line that is played slowly while notes are held out. Unlike the filter envelope, the amplifier envelope shape is fairly basic with attack and decay set as short as they can go and sustain and release set to full. Glide is set to about 20%, but can be fine-tuned to taste. Finally, a vibrato LFO is set up at around 10Hz with about 20% depth.

As was said above, simply adjusting the attack parameter of the filter envelope will result in drastically different sounds. In addition to this, adjusting the decay and sustain parameters of the filter envelope will adjust how the sound initially fades and remains while a key is being depressed. Changing the LFO to affect the filter cutoff will also allow for some cool, wah-wah-type effect.

Recipe 3: Effected, Evolving Pad

This patch is designed to be an out-of-the-ordinary pad sound. Being a pad, it is helpful to have some amount of polyphony available, but the patch sounds good monophonic as well. Due to a relatively fast LFO-controlling filter cutoff, the pad has a type or bubbly sound that makes it quite interesting. This type of pad would lend itself to most genres, but is mainly designed for an alternative pop genre.

Figure 1.17 Recipe 3—Effected, Evolving Pad.

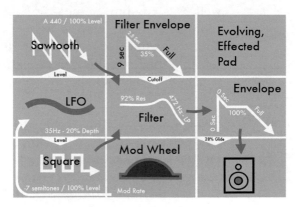

When re-creating this patch, the two oscillators are set to produce sawtooth and square waves respectively. The square wave oscillator is tuned to minus seven semitones from the sawtooth oscillator. The low pass filter's cutoff is set to 472Hz, while the resonance is almost fully engaged at 92%. The filter envelope's attack time is set extremely long at nine seconds, while the decay is set to 2.5 seconds. The sustain level is semi-low at 35% and the release is turned to full. This envelope shape allows for an extremely long filter evolution that will work hand in hand with the LFO. The amplifier envelope is fairly basic with attack and decay set to zero and the sustain set to 100%. The amplifier envelope's release is set to full. A fairly strong glide is set at 28%. Finally, a sine wave LFO is set fairly fast at 35Hz and is routed to the filter. The LFO in this patch is designed to constantly be engaged rather than controlled via a mod wheel. A 50% depth on the LFO is what was imagined for this patch, but it can be more or less based on your own personal preferences.

If desired, a third oscillator can be added either in tune with one of the others, or at another interval for a fuller sound. It is highly recommended that either pink or white noise is added at a low level to add some extra texture to the sound.

Recipe 4: Funk Lead

This Funk Lead patch is a fun lead sound that also works well as a rhythm or melody sound. The patch contains a chord-like sound with a tight, percussive filter envelope. Both the cutoff frequency of the filter and the resonance are set quite high resulting in an extremely bright sound. This patch will easily fit in a funk song, but will lend itself to any genre with a pop influence.

When re-creating this patch, the two oscillators are set to produce sawtooth waves with the second oscillator tuned up five semitones from the first. Both oscillators should be in a mid-to-high range, so 8' or 4' ranges are recommended. The low pass filter cutoff is set to 7200Hz with resonance at

Figure 1.18 Recipe 4—Funk Lead.

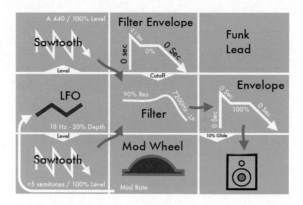

80–100% (just under the point of self-oscillation). The filter envelope's attack time is set to zero while the decay time is set to 0.1 seconds. The sustain and release controls are both set as low as they go. This envelope shape will create an animated percussive sound that bodes well for the rhythmic aspects needed for fun music. The amplifier envelope is a basic "on" shape with attack, decay, and release set to zero while sustain is set at 100%. A slight glide is set at 10%, and a relatively quick LFO, set to 18Hz, is routed to the oscillators pitch and is controlled via modulation wheel.

If a thicker sound is desired, a third oscillator producing a square wave and tuned to the first oscillator can be added. White or pink noise can be added in a small amount to add extra punch to the sound. If you find that the envelope is too percussive for your taste, increasing the filter envelope's decay time to around two seconds should suffice.

Recipe 5: Thick Bass

This bass patch differs from the first bass patch in that more emphasis is put onto the sound's "oomph." This bass patch is extremely thick and will be sure to rumble any subwoofer. As this is a fairly straightforward synth bass sound, it will fit nicely in any genre of music.

Figure 1.19 Recipe 5—Thick Bass.

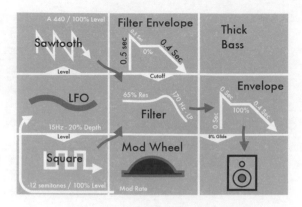

When re-creating this patch, one oscillator is set to a saw-tooth wave, while the second is set to a square wave that is an octave lower than the sawtooth. Since this is a bass patch, a low range should be set with the first oscillator at 16' and second oscillator at 32'. The lowpass filter cutoff should be set to 170Hz with resonance set fairly high at 65%. Both the filter envelope's attack and decay times are set to 0.5 seconds while the release is set to 0.4 seconds. Finally, the sustain is set to 0%. The amplifier envelope features the same "on" setting found in some of the previous patches with attack and decay set to zero and sustain set to 100%. The amplifier envelope's release, however, is set to 0.4 seconds. The glide should be set at around 8% and finally a 15Hz LFO should be routed to the oscillators' pitch for vibrato when engaged by the modula-tion wheel.

A third oscillator set to sine or triangle in the range of the first oscillator can be added for extra low end. Likewise, white or pink noise can be added to add extra grit to the sound.

Recipe 6: Percussive Staccato Pad

The sixth patch in our list is sort of unique in that it is designed to be a pad-like sound played on a monophonic synthesizer. That being said, this patch will also sound great

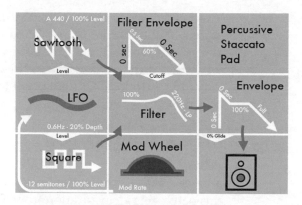

Figure 1.20 Recipe 6— Percussive Staccato Pad.

on a polyphonic synthesizer, it's just designed in such a way to make the most out of a monophonic synthesizer. The way the two envelopes are set up allows for a short, percussive sound, which then fades into a held-out, dark sound. This patch is not geared toward any type of genre and will lend itself well to any musical setting.

When re-creating this patch, the two oscillators are set to sawtooth and square with the square oscillator tuned down a full octave from the sawtooth oscillator. The low pass filter's cutoff is set to 220Hz, while resonance is fully cranked to 100%. The resonance can be toned down a bit if you do not like the self-oscillation sound present from full resonance. The filter envelope's attack is set to zero seconds, while the decay is set to about half a second. The sustain is left at 60% and, finally, the release is turned fully down to zero seconds. This envelope shape will yield a percussive sound that does not leave the filter fully closed. The amplifier envelope's attack and decay times should both be set to zero seconds as well. Both the sustain and release parameters are turned fully up. With such a short filter envelope and long amplifier envelope, the sound will be piercing and bright at its onset and then quickly become dark and ring out for a long time.

This patch was designed with no glide, but it can certainly be added. The LFO should also be tuned to taste. Another idea is a slow filter LFO that can move the filter open and closed while the sound is ringing out. If you want a weirder sound, a fast LFO around 35 or 40Hz controlling the filter could be added in via modulation wheel.

Recipe 7: 60s Organ

Although this patch is named 60s Organ, it is not meant to be a patch that faithfully emulates any type of organ. Instead, this patch gets its inspiration from 60s organs such as the Vox Continental.

When re-creating this patch, both oscillators are set to triangle waves with the second oscillator tuned an octave below the first. Pink or white noise should be added in a small amount in order to mimic the noise present in these old organs. The low pass filter's cutoff is set at 2800Hz with full resonance. The filter envelope's attack and release times are set to zero seconds while the decay is set at 0.16 seconds. Finally, the sustain is set at 34%. The amplifier envelope's attack and decay should both be set to zero seconds, while the sustain is set at 100%. Finally the release should be set around 0.17 seconds. No glide is required for this patch. If available, a 1Hz LFO should be routed to the amplifier in order to generate a slow,

Figure 1.21 Recipe 7—60s Organ.

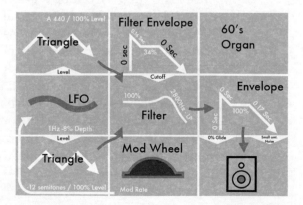

tremolo effect. If your LFO cannot be routed to your synthesizer's amplifier, routing the same 1Hz LFO to the filter's cutoff frequency is a good alternative.

This is a fun patch that can lend itself to any genre. Alternative or pop is most recommended, but it will fit nicely into any mix. The patch was designed with a polyphonic synthesizer in mind, but melodies and lead lines can be well executed on a monophonic synthesizer with this patch.

Recipe 8: Slow, Effected Lead

The eighth patch in this list is another take on a lead sound that is different from the stereotypical lead sounds out there. This patch contains an evolving filter with high resonance, which creates an effected sound that would be a nice addition to any piece of music.

When re-creating this patch, the two oscillators are set to square and sawtooth respectively, with the sawtooth oscillator tuned an octave above the square oscillator. The low pass filter's cutoff is set to 90Hz while, the resonance is cranked as high as it will go. The filter envelope's attack and decay are both set to zero seconds, while the sustain and release are both set to 100%. Both the amplifier envelope's attack and decay parameters are set to zero seconds, while the sustain and release are set to 100%. Glide is added at 30%. A 0.6Hz

Figure 1.22 Recipe 8—Slow, Effected Lead.

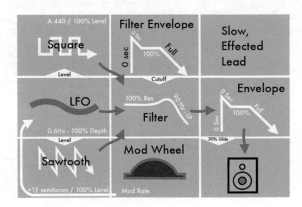

sine wave LFO is routed to the filter cutoff with 100% depth to slowly open and close the filter as you play.

An additional oscillator set to either sawtooth or square would be a good addition if you wish to beef up the sound. If your synthesizer has the capability of oscillator sync, engaging it on either oscillator would allow you to add some interesting harmonics to the sound.

Recipe 9: Chorded Trumpet

The ninth patch in our list draws its inspiration from a trumpet section, but is obviously designed so that it sounds more like a Roland Jupiter or Juno horn sound. The patch features two oscillators tuned at an interval in order to create a chord like sound.

When re-creating this patch, both oscillators are set to sawtooth waves with the second oscillator tuned up seven semitones from the first oscillator. The low pass filter's cutoff is set to 156Hz with the resonance at 64%. The filter envelope's attack time is set to 5.5 seconds, while its decay is set to 1.7 seconds. The sustain is set at 18% and the release is set at 0.05 seconds. The amplifier envelope's attack and decay are both set at zero seconds, while the sustain is turned up to 100%. Finally, the amplifier envelope's release is set at one second. No glide or noise is present in this patch, but can certainly

Figure 1.23 Recipe 9—Chorded Trumpet.

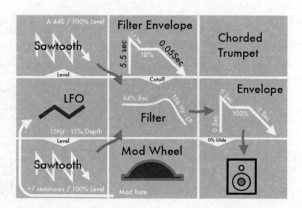

be added if desired. The LFO is set fairly standard at around 15Hz and routed to pitch for a vibrato effect.

Recipe 10: Wall of Sound

The final patch in our list is a great patch that can really thicken up an entire mix. The sound is massive and can be thought of in the same light as a huge power chord played on a guitar. This particular patch is perfect for industrial, hard rock, metal, or EDM music.

When re-creating this patch, both oscillators are set to square waves with the second oscillator tuned a full octave below the first. The low pass filter's cutoff is set at 2600Hz with resonance turned up to 65%. The filter envelope's attack is set to zero seconds with the decay at 0.25 seconds. The sustain is set at 20%, while the release is set to 0.38 seconds. The amplifier envelope's attack and decay times are both set to zero seconds, while its sustain is set at a full 100%. Finally, the amplifier envelope's release time is set identical to the filter envelopes release time at 0.38 seconds. Glide is set to 20% and a 15Hz LFO can be routed to either the filter cutoff or pitch.

A third oscillator set to a sawtooth wave and tuned identically to the second oscillator so that it's a full octave below

Figure 1.24 Recipe 10—Wall of Sound.

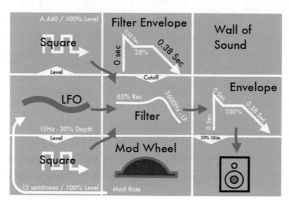

the first oscillator is a great addition to this sound. Additionally, white or pink noise can be added in order to introduce more grit to the sound.

Historical Perspective on Subtractive Synthesis

Subtractive synthesis can be traced back to the dawn of the electric age. The first leaps forward in electronic musical instruments often happened by accident when inventors were experimenting with electricity. One of the first instances of electronic tone generation came in the form of the oscillating light arc invented by Elihu Thomson in the late 1800s. Although not intended as an electronic tone generator, Thomson found that his light arc would hum when operated.[1] William Duddell, another early inventor, took Thomson's design a bit further by wiring many oscillating light arcs together in order to perform "God Save the Queen."[2] Duddell's performance on the oscillating light arc is often considered to be the first performance of an electronic instrument. In essence, what Thomson, and later Duddell, invented was the world's first oscillator.

Early Electronic Instruments

Not long after the oscillating light arc, inventors such as Thaddeus Cahill and Leon Theremin began inventing electronic devices designed to play music. Cahill's Teleharmonium and Theremin's Etherphone (later named Theremin) were among the first electronic instruments ever designed.[3] The Theremin went on to achieve success influencing future electronic music inventors. The Theremin, which is still in production today by numerous companies, is a noncontact, electro-harmonic instrument. The Theremin features two antennas (one for pitch, one for volume) that the user places his or her hands between in order to control the instrument. Two internal oscillators then produce one continuous wide range pitch via beat frequencies.[4] The Theremin was widely

Figure 1.25 A modern Moog Theremin; a direct ancestor to the original Theremin. Photo courtesy of www.switchedonaustin.com.

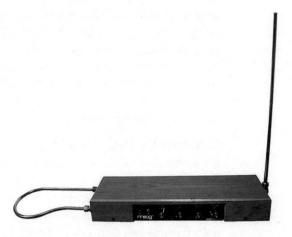

used in virtuoso performances as well as a wide array of early science fiction movies.

Soon after the Theremin's debut, a French inventor by the name of Maurice Martenot released a similar sounding instrument known as the Ondes Martenot.[5] The Ondes Martenot utilized a similar principle as the Theremin, but a keyboard and suspended ring was used as a means of control rather than antennas. Users would place their finger in a small, suspended ring underneath the keyboard and move their hand up and down the length of the keyboard in order to control pitch in a continuous manner.[6] Due to the fact that a keyboard was available to be used as a reference, users could create stable pitches much easier with the Ondes Martenot than with the Theremin. Finally, the Ondes Martenot featured a physical block that the user would depress in order to change the amplitude of the instrument.

In the late 1930s, a new electronic musical instrument, designed by John M. Hanert, Laurens Hammond, and C.N. Williams, would be introduced that would revolutionize electronic instruments. This instrument was known as the Hammond Novachord. Not only was the Hammond Novachord able to electronically generate tones, it was able to do so with full polyphony, meaning that each key on the keyboard could be played simultaneously. Full polyphony was so revolutionary that it wouldn't be commonplace on synthesizers until the late 1970s and 1980s. The Novachord was also revolutionary in the fact that it is considered to be the first synthesizer to utilize envelope generators.[7]

Soon after the Hammond Novachord was released, a new synthesizer called the Ondioline made its debut. The Ondioline, invented by Georges Jenny in 1941, was unique in that not only did it offer envelope generator control and a complex filter bank, it was one of the first synthesizers to feature a wealth of adjustable parameters that could easily be recalled, resulting in the first successful synthesizer to utilize a primitive form of preset control.[8] Thanks to its complex filter bank and advanced envelope contours, the Ondioline was especially good at re-creating various orchestral sounds like oboes and violins.

As early electronic instruments like the Theremin, Ondes Martenot, and Odioline began to make an impression on the music industry, many influential people were taking notice. Two such people, Herbert Belar and Harry Olson, who were working for RCA, began work on their own electronic musical instrument. The instruments Olson and Belar would create would end up being known as the first fully programmable synthesizers—the RCA Electronic Sound Synthesizer MK I & II. The RCA synthesizers featured 12 and 24 oscillators respectively as well as a wide array of envelope and filter control. Perhaps what was most revolutionary about the RCA synthesizers was that they were fully controllable via an early binary computer that acted as a sequencer. The user would program the synthesizer by not only adjusting the machine itself, but by creating a sequence on perforated paper that would then be fed into the binary computer, effectively allowing the synthesizer to play itself. Although the RCA synthesizers were never produced commercially, they have gone down in history as the direct ancestors of the synthesizers we know and cherish today.

Moog and Buchla Modular Systems

In the early 1960s, two synthesizers would emerge from separate ends of the country almost simultaneously that would change the face of synthesis forever. These instruments were the Buchla 100 Series Modular Electronic Music System and the R.A. Moog Modular System. Many heated arguments have arisen as to whether Moog or Buchla created the first voltage controlled analog subtractive modular synthesizer. It is hard to determine who in fact was first, since they both emerged around the same year—1964. It is undeniable that the Moog went on to have more success, but the Buchla introduced things to synthesis that are still utilized today.

Dr. Robert Moog got his start in electronic musical instruments by building Theremin kits. Moog's love for the Theremin not only got him interested in electronic musical instruments, but would introduce him to early electronic musician Herb Deutsch. Deutsch began to work with Moog on creating an electronic instrument that was desperately wanted by electronic musicians. Moog began creating a series of voltage controlled modules that each provided a different, sound-modifying function. Quickly thereafter, Moog began placing each of these modules in a wooden enclosure and selling them to the public. The modules Moog created varied from oscillators and LFOs, to envelope generators, attenuators, and amplifiers, and to filters and mixers. The modules would then be physically connected together via quarter-inch

Figure 1.26 The first Moog modular system commercially produced. The instrument is part of the Stearns collection at the University of Michigan.

patch cables in order to build a desired sound. Moog is often credited with introducing the traditional keyboard as a means of designating pitch. Although using a keyboard to control a synthesizer was controversial among early, electronic music pioneers such as Morton Subotnick and Vladimir Ussachevsky, the idea caught on and has become the standard means of control among synthesizers.[9]

While Moog and Deutsch were creating what would become the Moog Modular in New York, Don Buchla was working on a modular synthesizer of his own near San Francisco, California. Buchla was approached by Morton Subotnick, along with a variety of other experimental electronic music artists working out of the San Francisco Tape Music Center, in order to build an electronic musical instrument. Subotnick made very clear that he did not want the synthesizer to be controlled via a traditional keyboard as he thought it would trap him into making tonal music, something he did not want to do. A few months later, Buchla arrived at the San Francisco Tape Music Center with a prototype of his modular system. Unlike the Moog, the Buchla synthesizer utilized a combination of touch plates and sequencers in order to designate pitch. In fact, the Buchla sequencer proved so successful, Moog ended up

Figure 1.27 A 1969/70 Buchla 100 Series manufactured by CBS Musical Instruments. Image courtesy of Rick Smith with www.electricmusicbox.com.

producing a similar sequencer for his modular system years later. Buchla's approach to synthesis was quite different than Moog's, and the term West Coast and East Coast synthesis has been coined for Buchla and Moog's synths respectively.[10]

As stated above, the Moog went on to monumental success and began showing up in a number of studios and musician's homes despite the enormous cost. One such adopter of the Moog modular—Walter (now Wendy) Carlos—created an album featuring reworkings of famous Bach pieces on the synthesizer. This album, "Switched on Bach," is responsible for bringing the synthesizer out of the experimental music studios and into popular culture.[11]

Analog Mono Synths of the 1970s

With the success of the Moog modular, synthesizers started to become a hot commodity. Due to the high cost and steep learning curve, though, many musicians were begging for something smaller, cheaper, and easier to use. Answering their call, Moog released the MiniMoog in 1974. The Mini-Moog was a compact, road-ready synthesizer and keyboard combo that was internally wired, which negated the necessity of using patch cables to produce sound. For the first time,

Figure 1.28 Arp Odyssey. Photo courtesy of www. switchedonaustin.com.

the synthesizer became something practically anybody could learn and was small enough to be brought on tour. The Mini-Moog revolutionized synthesis and, soon, a wealth of companies began releasing compact analog synthesizers. ARP, Yamaha, Korg, Roland, EMS, Crumar, EML, Oberheim, and Sequential circuits were among the vast number of companies that produced analog mono synths that would create a paradigm shift in the music industry.

The Digital Age and Beyond

As analog subtractive synthesizers became commonplace amongst musicians, companies began producing digital synthesizers at a lower cost. Musicians began flocking to these new digital synthesizers due to their low cost, the fact they stayed reliably in tune, and because they featured presets making large, time-consuming tweaking unnecessary. As more and more musicians began selling off their old analog subtractive synthesizers for more complex digital synths, subtractive synthesis seemed all but doomed. The mid 1990s, however, brought with it a small resurgence of analog interest. At first, this new appreciation for analog subtractive synthesizers seemed to only have a cult following, but it was soon determined that users wanted new, subtractive synthesizers. Companies began producing software emulations of old analog gear as well as a number of digital subtractive synthesizers. As demand grew, companies like Moog and Dave

Figure 1.29 The Korg Little-Bits synthesizer system is a modular DIY analog synthesizer, which gives users an insight into the electronics and mechanics present in analog synthesizers.

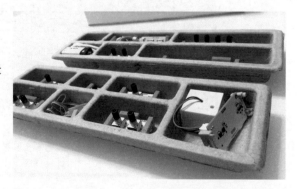

Smith of Sequential Circuits began producing all-new analog subtractive synthesizers. In an almost ironic turn of fate, the synthesizer market was once again being flooded with not only analog subtractive synthesizers, but with modular analog synthesizers as well. The synthesizer industry has effectively come a full 360 degrees from where it started, and suffice it say, it is an exciting time for synthesis.

Notes

1. Mary Jenkins, *Analog Synthesizers*, Focal Press, 2007, p. 20.
2. Jenkins, p. 63.
3. Jenkins, p. 24.
4. Vasilyey Yurii, "History and Design of Russian Electro-musical Instrument 'Theremin.'" Paper presented at the annual Audio Engineering Society Convention. May 20–23, 2006, Paris, France.
5. Jenkins, p. 31.
6. "Ondes Martenot," *Encyclopedia Britannica*. Retrieved from http://www.britannica.com/EBchecked/topic/428993/ondes-martenot
7. Michael Murphy and Eric Kupp, "An Examination of Early Analog and Digital Sampling—The Robb Wave Organ Circa 1927." Paper presented at the annual Audio Engineering Society Convention. May 4–7, 2013, Rome, Italy.
8. Carlton Garner, "Electronic Instrument," *Encyclopedia Britannica*. Retrieved from www.britannica.com/EBchecked/topic/183802/electronic-instrument/53835/The-tape-recorder-as-a-musical-tool#ref111938
9. *Moog*, DVD. Directed by Hans Fjellestad, 2004.
10. *I Dream of Wires*, Blu-Ray. Directed by Robert Fantinatto, 2013.
11. Robert A. Moog, "Electronic Music," *Journal of the Audio Engineering Society* vol. 25, no. 10/11 (November 1, 1997), pp. 855–861.

SAMPLING AND SAMPLE-BASED SYNTHESIS

2

Synthesizers changed music forever, and even though they are capable of a wide variety of sounds, designers have continuously sought to emulate acoustic instruments. Yamaha's DX-7 was one of the first synths that came close, and it did what no other instrument had done before; offer an affordable instrument that realistically mimicked other instruments. Once that door was slammed open, developers sought to do a better job of emulating traditional instruments with improving technology, and sample-based instruments started to take over.

Another draw to sampling is live performance where the efficiency of resources and access to a variety of instruments is paramount. Instead of bringing every single individual instrument on tour, a sampling setup allows you to have all of the sounds you want to use in a single place. There is something magical about using your actual vintage synthesizers on the road until something goes wrong or until you try to make quick patch changes between songs.

Sampling has its own history separate from synthesis and it is big enough to fill an entire book all by itself. Some of the history is explored later, but the bulk of this chapter is a look at the point where synthesis and sampling overlap. In spite of having substantially different sound engines, both synthesizers and samplers have similar modulation, envelope, and effects sections. These similarities mean that a sampler can

Figure 2.1 Sampling goes mobile on iOS with Garageband.

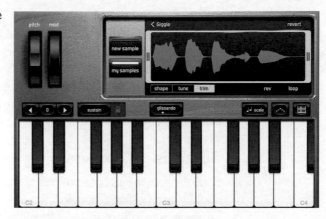

easily be used in the same way you would use many types of synthesizers, with very similar results.

Sampling is a tedious art form because it can take hours and hours to create a sampled instrument, but there is still an element of art and you have to become a craftsperson with a special attention to detail. The caution that you should keep in the forefront of your mind throughout this entire chapter is that you shouldn't get lost in the technical aspects of sampling. During the process of creating a sampled instrument, you should always ask yourself if the decisions you are making will make a more expressive instrument or just one that is accurate. Embrace the art over the science when it comes to sampling!

Basics of Sampling

The process of sampling has the following phases:

1. Source capture
2. Editing
3. Mapping
4. Tweaking

After covering the basics of sampling, there are several specific projects which are described in detail to showcase the power of collaboration between sampling and synthesis. The examples in this chapter are DIY instruments that take advantage of envelopes, filters, and other synthesis modules. Projects that you could do could also include nontraditional sound sources, which create synthesizers out of sounds from everyday things such as wind, shutting doors, or just about anything you can record.

Sampling Overview

Sampling is the process of mapping audio files to MIDI notes so they can be triggered for playback. The only real difference between sampling and synthesis is that sampling uses audio files as the sound source instead of generated sounds from oscillators or other tone generators. Digital synths such as the DX-7 mimicked instruments, such as brass or pianos, but a sampler uses actual sound files recorded from these instruments, and that means they will sound exactly like the actual instruments.

As computing technology improved and became more affordable, samplers and ROMplers (with no recording capabilities and only sample playback functionality) became more popular and overtook synths such as the DX-7. Samplers are as popular now as they have ever been and are a common production tool due to the power and flexibility of software offerings.

Figure 2.2 Editing samples in Logic's EXS24.

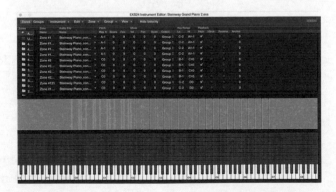

Every major, digital audio workstation has its own sampler, which can be used to create complex instruments, but there are couple samplers such as MachFive and Kontakt that have even more features and can be used with any DAW. Mach-Five in particular is a sampler that is of interest to synthesizer users because it has a series of synth engines that work with recorded samples to create amazing sounds. Examples of this include a four-operator FM synth, a granular synth, and a wavetable synth. In many ways, MachFive is the only instrument you'd ever need, but we all know that one is never going to be enough to satisfy our creative juices.

Another new trend in sampling involves mobile devices that can record, edit, and trigger samples using intuitive touch interfaces in a portable package. Very few mobile samplers are capable of complex instruments, but that is going to change and improve as mobile devices improve. For the purposes of this chapter, we are going to use the EXS24 for illustrations, due to the popularity of using Logic Pro X for this type of musical creation.

Source Capture

Capturing good sound sources is the most important part of sampling because nothing you do afterwards will be able to significantly change the original source material. Sampled instruments are created from the recordings made during the capture phase, and so it is critical that the sounds are recorded at the highest quality possible. There are three typical ways to record the source, which include using microphones, direct connections, and internal connections.

Sample Recording Considerations

Using a microphone in the sampling process is the traditional method for capturing acoustic instruments, but when it comes to synthesizers, microphones should only be used when no other option is available. Some instruments do not have output ports, but they do have built-in speakers that can

Figure 2.3 USB microphone from sE Electronics.

be recorded with microphones. Using microphones also creates a more complex situation because there are many different types of microphones, and preamps/converters are required.

If you decide to use a microphone in the sampling process, then you'll want to make sure to check a few things to maximize the quality of the audio recordings. The first is that you should listen to the output of the synth speaker to make sure there isn't too much noise. If the instrument has two speakers, then make sure that you are recording the one that sounds better and has less noise. You also want to make sure that there isn't any mechanical noise from the keys or from pedals. Many issues you can fix, but you should try to minimize the ones you can't.

The other consideration you need to think about is the acoustic space of the room where you are recording. This needs to be a quiet room and a room that sounds good because everything that is in the space will be recorded by the microphone. If you have to record in a nonrecording studio space, then make sure that you listen carefully before recording to see if there are any soft buzzes, hums, or other sounds coming from inside/outside the room. Don't compromise in making sure there are no noises because if there are, then it will affect your sounds in a negative way that will be very difficult to fix later.

The best way to check for audio quality while sampling is to put on a nice pair of headphones, and then listen critically to each sample. If you have access to a professional, quality studio, then using their speakers to monitor the sound is likely acceptable; but if there is any doubt, then use high quality headphones which can act as a microscope for a sonic analysis to ensure low noise and great sound.

While it is possible to use an inexpensive microphone and equipment set-up, this should be avoided because you will not be capturing the best possible version of the sound. The cheapest microphones have the potential to add noise into the signal at the very earliest stages, which might make your sampled instrument unusable.

Even though you don't need to worry about the acoustics of the room, you still need to listen for buzz and hum and other noises in the signal. Some vintage instruments are likely to have more noise than others and we will talk about some things you can do when we get to the section on editing.

Sample Recording Tools

As stated above, describing an entire microphone setup is outside the scope of this chapter, but here are some examples of equipment options to give you an idea of the possibilities. The three scenarios are a USB microphone, an all-in-one interface with a separate microphone, and a setup with all individual components.

USB Microphone

I use a sE Electronics 2200a USB microphone, which connects to a laptop or iPad and is an excellent choice because it is low noise and sounds great. The microphone has a built-in headphone jack, which means you can listen to what you are recording directly on the microphone. Plus you can use the microphone with other audio interfaces since it has a traditional XLR connector in addition to the USB port. This option is literally a complete studio that can be used for sampling.

Figure 2.4 USB connection and headphone port.

Audio Interface

There are many different audio interfaces available in the marketplace, and they come in a variety of configurations at a wide variety of prices. Assuming you are planning on sampling synthesizers, then you won't need an audio interface with a lot of inputs and can likely manage with one or two. Most interfaces connect with your computer via USB, Firewire, or ThunderBolt. After verifying compatibility with your computer and choosing the number of inputs, you'll want to verify it has microphone preamps and XLR inputs.

The choice of microphone is complex because there are many different types and brands, and, honestly, you may really want to use different microphones for different recording projects. The three primary types are dynamic, ribbon, and condenser. All three convert acoustic energy to electrical energy but do it in different ways that result in different sound characteristics. Dynamic microphones rely on a moving diaphragm in a magnetic field, which results in a less detailed sound with less clarity in the upper frequency range. The character of these is often great for electric guitars, live performance, or other percussive sources. Ribbon microphones rely on a delicate strip of metal in a magnetic field, which makes them among the most sensitive, but also the most expensive and prone to damage from rough handling and voltage from phantom power. The condenser type relies on the difference

Figure 2.5 Focusrite Saffire Pro 40 Firewire interface.

in distance between two charged metal plates and is very sensitive to small differences in acoustic energy, which translates to higher clarity in frequencies and sensitivity for sound at softer levels.

A condenser microphone is recommended for most synthesizer sampling, but any of the types could work. In addition to a microphone, you'll also need a microphone cable, which is typically a three-pin XLR cable. Condenser microphones also require power in the form of phantom power, which is available on most interfaces. This 48-volt power option is required for the microphone to work. Buying an interface and microphone can cost as little as $500–1,000 or be as expensive as $1,000–10,000.

Individual Components

This option is essentially the same as the one previous, but breaks the audio interface into its individual pieces. These include a microphone preamp, an analog to digital converter, and a computer interface. This chain has the potential to be the very highest quality, but is definitely the most expensive and most complex. If you are sampling synthesizers for your own use and for live performance, then the individual component option is likely overkill because the quality of the first two options are both acceptable, but if you want to sell sampled instrument banks, then a higher quality option is important.

Microphone Guidelines

It is possible to set your input levels too high so they distort or too low so additional noise is introduced, and so you want to set it at the highest level possible without distorting. If the source synth sounds best when turned up very loudly, then adjust the microphone preamp appropriately. In some cases, you'll have to use a pad on either the microphone or on the interface, which turns down the input by a set amount.

Direct Connections

Recording a synthesizer with the direct connection is the ideal way because it removes acoustics and room considerations from the process. While you still need an audio interface for your computer, you'll connect the instrument with cables instead of using microphones. While this is the optimal method for recording from a synthesizer instrument, it is just as complicated as any of the other methods and so there are a few things we need to cover to help you be prepared.

Matching Outputs

The first step in recording samples using direct connections is to match the number of outputs on your instrument with the number inputs on your audio interface. In most cases, you'll be dealing with mono or stereo outputs, but in some cases, you'll have four, six, or eight outputs. If you want to be able to record all outputs simultaneously, then you need to have an audio interface that matches. It is unlikely that you'll need multiple outputs, but some synths have alternate outputs for layering patches and a few instruments have surround outputs, which can all be sampled into a single patch.

In addition to matching the right number of outputs, you also have to match the type of output. If the instrument uses 1/4" unbalanced cables, then you'll need to make sure the interface can accept an input with that type. If the instrument has a digital audio output, then having an interface that can accept that format is critical. Typical connections include the following:

1/4"—Balanced (TRS) and unbalanced (TS)
1/8"—Balanced (TRS)
XLR
Optical
Coaxial / RCA

Some audio interfaces have combination connectors which accept both 1/4" and XLR cables, but you'll still need to choose between a mic level input and a line level input. One important troubleshooting tip is to make sure the synth is working and sounding like it should in your audio interface by plugging in headphones into the headphone jack of the instrument to make sure that the sound of the synth is there and matches what you are expecting.

Once everything is connected, the next part of the recording process is to set the levels appropriately. Nearly every interface has level adjustments that make it possible to turn up or down the input levels. If these are set too low, then there is a chance that there will be too much noise in comparison to the signal in the recording; but if they are set too high, then you will face distortion. Distortion is worse, but both are a problem when creating high-quality sampled instruments.

Recording Notes and Layers

To create realistic-sounding instruments, you need to record as many notes as possible and, when appropriate, at different velocities. Some synthesizers don't have velocity-sensitive capabilities and so it makes sampling of them very easy, but when a synthesizer has variable velocity or any modulation controlled by key velocity, then sampling becomes a more intricate process. If the synthesizer sounds different when you play the keys with more or less velocity, then recording at multiple levels is important. Listen closely to the synthesizer when you play it at different levels because if the loud sounds are just turned up, but are no different than the soft ones, then you don't need to record alternate velocities. Samplers can easily play recorded samples at different levels, but it's the change in sound across the velocity spectrum that can't be replicated without recording additional samples. If the instrument has MIDI or C.V. input, then sequencing the notes for sampling can often help ensure an even velocity spread for samples.

Figure 2.6 Velocity layers example.

Editing

The editing process of your source material involves removing empty space, making sure the levels are correct, removing noise, and anything else needed to make sure the samples sound good. If you recorded all of the notes for one set of the instruments into a single file, then start by processing the audio with everything that needs be done to all of the notes at once. This includes adjusting the level globally, performing noise reduction if necessary, and adding any additional audio effects if desired.

Once you're satisfied with the entire file, then you begin to split all of the notes into individual regions. Most digital audio workstations have a feature called Strip Silence, which removes parts of the audio file that fall below a certain threshold, and this is very useful when working with this type of file because it will save time by splitting the files up automatically. It is important to look at the beginning and end of every file while zoomed in to make sure the start point is at the very beginning of the audio portion of the file and at a zero crossing point, which means that the waveform starts at the exact point where the audio is crossing the axes instead of partway through a wave (this is to prevent a popping

Figure 2.7 Editing individual files in the Logic File Editor.

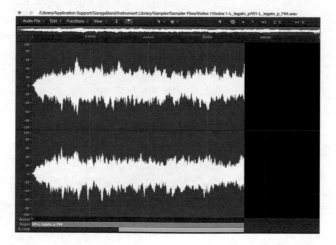

sound). Use fades to smooth the start or end of each file if needed.

The next step is to have each sample converted to its own file, which is important for the sake of organization and to keep things simple for the sampler. Audio software references the original long audio file, and even when it looks like they are separate files, there is a good chance that they are still all one big file. Converting clips to new files is especially important if you are using a sampler that is a part of your editor, such as the sampler in Logic Pro X, because you can drag the sample files into the sampler and it still won't use individual files. The best option is to export all of the samples as individual files into a folder with descriptive filenames that are numbered sequentially and then import them into your sampler. While you can use any naming convention you want, it is easiest when you use a name that includes the note name, note octave, and some indication of the patch from the original instrument. Double check each file by opening them to make sure they are the appropriate length and the correct note.

Mapping

The mapping process takes each individual sample and assigns them to MIDI notes. Sample files are loaded into zones, which tell the sampler the root note, the key range, and the velocity level. If you recorded a sample for each note, then each zone will cover a single note, but if you recorded a sample with the goal of covering multiple notes, then you'll need to adjust the zone range to match the desired range. All zones require a root note setting, which tells the sampler what transposition to use for the sample. The root note

Figure 2.8 Mapped zones.

Zone	Pitch			Mixer					Key Range		Playback					Vel. Range			Sample			Loop						Audio File
Name	Key ▲	Coars	Fine	Vol	Pan	Scale	Output		Lo	Hi	Pitch	1Shot	Reverse	Anchor		On	Lo	Hi	Start	End	Fade	On	Start	End	Tune	XFad	E.	Name
Zone #302	G#2	0	0	0	0	0	Group ⇕		A-1	G2	✔	▢		▢	0	▢	0	127	0	169561	0	✔	78049	169561	0	0	▢	

Figure 2.9 Zone settings.

should correspond to the pitch of the sample and many samplers will set this automatically based on the sample name or analyzed pitch upon import.

The next task in mapping is to set the velocity layers, but only on instruments where you originally recorded multiple levels of velocity. Due to the limitations of MIDI, there are only 128 steps of velocity, but there are no guidelines on where to set the velocity ranges for the samples you've recorded. The process for setting velocity ranges is part common sense and part trial and error, and the only way to confirm that it is set correctly is by testing it out to see if it sounds right. In addition to playing the instrument using a controller keyboard, you could also use a MIDI test file, which plays through all notes and ranges so you can hear how consistent it is in each of them.

Loops

Looping is another important task that takes place inside the sampler and unless your sampler has an auto loop function, this is something that can potentially take a very long time. Not all instruments require looping, but when working with synthesizers, looping often makes the most sense. The best approach when setting loop points is to be very methodical and pay close attention to the details. Following are a few steps to follow in the looping process.

Find an Optimal Loop Section

A looped section is repeated over and over for the purpose of sustain and so it has to be a section where the beginning and end are similar in level and timbre, otherwise there will be a bump in the sound when the note is sustained. If you are sampling using the waveform option, then the entire file is typically looped, which makes this very easy, but you'll still

Figure 2.10 Looping at the zero crossing point.

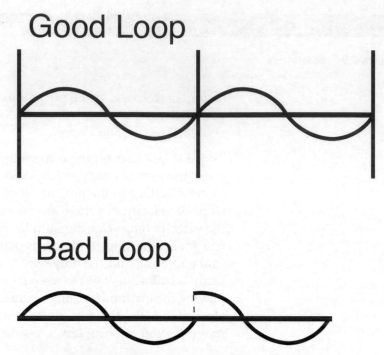

Good Loop

Bad Loop

need to check each zone. Instruments that change over time with either envelopes or filters may be more difficult to loop, but as long as you can find even a short section to loop, then it will work. Some loop points will be short and some will be long, and this depends on the type of sound of the samples. Instruments with a very consistent sound will often work well with short loop points while some instruments benefit from longer loop points.

Choose the Loop Direction

Typical options for loop direction include forward, backward, and forward/backward. Forward is the most common and, after it plays the loop section once, it goes back to the beginning and plays it over and over while the note is sustained. Forward/backward is a good option when you aren't able to make the forward loop work, because there is never a transition bump in level since the loop playback bounces forward

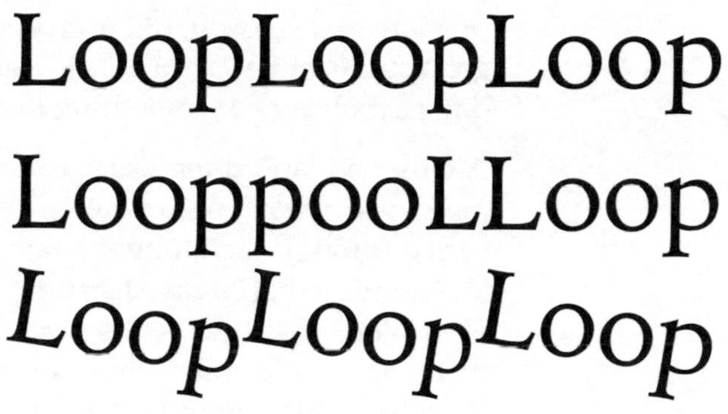

Figure 2.11 Typical looping options.

LoopLoopLoop

LooppooLLoop

LoopLoopLoop

and backward over and over. The main issue with that is that the bouncing back and forth is often quite audible.

Set Crossfades

In addition to setting your loop points at zero crossings, it is an option to use fades that cross over the loop point. Crossfades create overlap so that, as one sound ends, the level goes down just as the level comes up from the beginning of the loop. If you aren't able to create a loop that sounds smooth as it continuously loops, then try a crossfade to smooth it out. The length of the crossfade is less important than making it sound good, so start with a short one and increase it until it sounds just right. The best crossfade is one that you can't hear.

Adjust the Release

Samplers handle the release section in a variety of ways, which include fading out in the loop section or continuing in the sample after the loop section. Sometimes this is called a release and other times it is called a note-off trigger. It is unlikely that you'll have that much control over this, but you will want to test it to make sure that the release settings work for your instrument.

Advanced Features

The rest of the mapping process involves using the advanced features of your sampler. Many samplers let you use key-switching, which means that you'll map all of the zones as previously discussed, but each key-switch layer needs a trigger. You can either use a note lower on the keyboard to trigger different players or something like the mod wheel.

Another advanced feature is round-robin, which uses multiple zones in the same key range and velocity range, which are cycled through each time the same note is played. This mimics the way that many instruments make different sounds, even when the exact same note is played over and over. If you want to use this technique, then it easily means two or three times the amount of recording.

Scripting is perhaps the most advanced feature and only exists in the most advanced samplers, such as Kontakt from Native Instruments. Scripting relies on computer-style programming to create additional parameters for mapping and performance. An example is a script that recognizes a played chord and then offsets the timing of each note to mimic the strum of a guitar. It could also translate the MIDI input performance of a piano-style keyboard so that it sounds more like a nonkeyboard instrument, such as a violin or trumpet. This could affect timing and sample selection so that it sounds and feels like the real instrument. Scripting is the most advanced feature and is complicated enough that most mere mortals aren't able to program much, but these same samplers almost always come with sampled instruments that take advantage of this amazing technology.

Figure 2.12 The advanced features of Logic's EXS24 are just as powerful as any other, but perhaps the hardest to use because the interface is difficult to understand and use.

Figure 2.13 A MIDI test file triggers the samples at different velocities across the entire range. Adjust the tempo to try them at different speeds.

Tweaking

Once you have created your instrument by recording the source, editing the samples, and mapping them across the keyboard, then it's time to put it through its paces and begin the tweaking process. It is unlikely that you will create the perfect instrument in your first try and it is more likely that you'll start with something fairly close and then spend a substantial amount of time adjusting velocities, zone ranges, and other parameters. Every instrument is different and you will need to make yours work through a lot of experimentation.

The process of testing the completed instruments involves playing them with MIDI controllers, using any MIDI files you have to test them out, and having a bunch of other people test them out too. While tweaking the instruments is one of the most rewarding parts of the process because it all comes together as a useable instrument, it is important to pay attention to the smallest details and not lose sight of the goal of creating a realistic instrument that sounds great and is very playable.

Sampling Examples

The following three scenarios showcase typical situations that you face when sampling synthesizers.

1. Waveform sampling
2. Basic instrument
3. Advanced instrument

Waveform Sampling Example

Recording the waveforms of a synthesizer is the most basic way to create a sampled synthesizer, even though it isn't usually the oscillators that give a synthesizer its distinctive sound. Most samplers, however, have built-in filters and modulators, so you can use the basic waveforms from an instrument and then use the rest of the sampler to create a synthesis experience. When sampling the basic waveforms, you can use as little as one cycle of the oscillator or much longer segments in the cases of the analog synthesizer that may have pitch drift. When there is drift present, it can add to the authentic sound of the sampled instrument. You'll have to use your best judgment when deciding between the two, but both will work.

The first step is to set the synthesizer so that the oscillator is reaching the outputs unprocessed, which means turning off all filters, envelopes, modulators, or other effects. Listen to the synthesizer in different ranges to see how consistent the oscillator plays across the entire range and this will help to determine if you need to record the oscillator on every note or if you can record it once every three or four notes. Samplers use a time function to transpose notes when using a limited number of samples, which means you can have a single root note be triggered for multiple notes on the keyboard and each note that isn't the same pitch as the root note will be transposed up or down.

When recording the waveforms, you can either start a new recording for each note or you can record all of the notes with pauses in between them into a single file. Some tasks like noise reduction or adjusting gain can more easily be accomplished when all of the waveforms are in the same file, but it

Figure 2.14 Single cycle of a sine wave.

One waveform cycle

comes down to a personal preference and choosing a work-flow that makes sense to you.

Basic Instrument Example

In this scenario, you'll create a patch on your synthesizer that has the oscillators active, but also runs through filters and other modulators. The basic sound can include motion with filters and envelopes, but should not involve more advanced motion, which relies on the modulation wheel or after-touch. Even though some of the most interesting parts of using synthesizers live are real-time control of filters and other modulation effects, for this scenario, you want to create a complete patch, but still keep it simpler than the full potential of a synthesizer.

When creating a sampled instrument using simple wave-forms, the sampler has to loop each waveform to create sustained sound. In this scenario, if you are using a patch with any motion in the filters or with the envelope, then you'll have to record the entire length of the sound until it reaches a sustain point or until it dies out. Using the loop feature allows

Figure 2.15 Basic instrument example recipe.

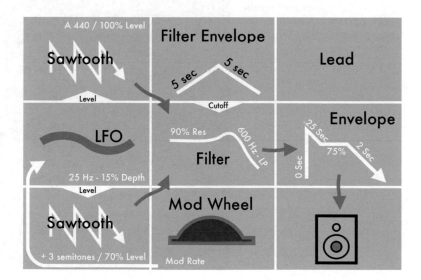

you to sustain a note even if it isn't recorded at the full length, but if the patch you use doesn't have a suitable looping point, then you have to record the full length of the patch during the sampling process.

A finished instrument of this type sounds exactly like the original synthesizer patch, but you won't have any control over the sound in the same way that you would when using the original synthesizer.

Advanced Instrument Example

The first two scenarios are relatively easy to create, but with the third scenario, there are a lot more elements to manage and the final instrument can be quite unwieldy. The goal with this instrument is to capture the same performance ability that you have with the actual synthesizer. This is possible by using advanced sampler capabilities that aren't available with every sampler, but if you are serious about sampling, then you should invest in any one of the most powerful samplers.

Figure 2.16 Key-switching with EastWest. Different shade of gray on keyboard represent notes that switch between sample sets.

The primary feature that you can use to create more performance control is called key-switching; this allows you to create multiple versions of the same instrument, which are selected by using an additional controller such as the mod wheel or a lower key on the keyboard. Several samplers allow you to crossfade between these instruments as you adjust the mod wheel, which creates a similar effect to the actual instrument as you move its mod wheel.

The recording process for this scenario involves recording all of the notes and velocities for each of the settings you want to be involved with the key-switch. An example of this would be recording a variety of sounds with a filter cutoff frequency at different settings or with different attack times in the primary envelope. The amount of work required in the recording process in this scenario is substantially more than with the two previous scenarios because of the sheer amount of audio you have to record. Keep things organized because in the editing and mapping process, things become more complicated.

Synthification

Samplers fit as a single piece of a much bigger ecosystem of digital audio tools, processors, and effects. It is possible to use the default filters and effects in your sampler of choice and while they may be excellent, it is possible to use external effects as a part of the sound. Combining elements in your DAW creates a modular synth type experience that creates a more powerful experience than a sampler is capable of by itself.

There are several tools that allow you to emulate individual synth modules, such as filters and effects, and these can be powerful additions to the realism provided in the sampling process. The primary option is any Convolution Reverb, which has the ability to create impulses. Impulse response (IR) technology is able to send a sweeping sine wave through a component and then extract spectrum information from the result. The resulting IR can be applied to other audio signals to change their spectral content to match the same effect as if it was being processed by the original module. The primary limitation is that the IR is a snapshot of the spectral content and isn't able to dynamically change based on the input signal. If the input has a loud bass component that is slightly distorting, the IR process will not react in the same way and will not have the same results.

Auto-Samplers

Auto-samplers are capable of creating fully finished sampled instruments without needing to manually edit, loop, or anything. The process works by sending MIDI data to a destination instrument, recording the audio output, and then syncing the MIDI information with the audio material. The parameters of the instrument are established before beginning the sampling process, and the application does all of the rest. Auto-samplers can save an enormous amount of time by finding and optimizing loop points, setting velocity layers, creating multiple layers for key-switching, and even tuning out of tune notes.

Figure 2.17 Ever since Apple closed Redmatica, Sample Robot has been one of the few companies interested in auto-sampling.

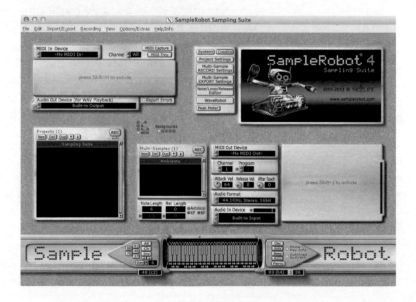

All instruments with MIDI inputs and audio outputs can be auto-sampled, but also instruments with C.V. inputs will work if you have a MIDI-to-C.V. converter. The best auto-sampler, called Keymap Pro, was bought by Apple in 2010 and closed down permanently. Another company has stepped up to the plate with Sample Robot, and it is capable of creating very functional instruments. An example with Sample Robot is included here to showcase the power of the process and, afterwards, you will likely wonder why anyone would ever take the time to manually sample these types of instruments. The simple answer is they shouldn't, but the art of sampling lives on with acoustic instruments that can't be triggered by MIDI.

Sample Robot Example with Vermona Analog Drum Synthesizer

The following images highlight an auto-sampling project with an analog drum synth from Vermona. The reason that this is an excellent instrument to sample is because, as an instrument, it doesn't have presets; so if you create a set of percussion instruments that you like, then you have to either make detailed notes or use a camera to remember the settings. The sampling process lets you create a digital version of the synth patches that you can use any time you want without having to adjust all of the knobs. The actual instrument is far more fun to use, but there are many times I am glad to have the sampled version.

Figure 2.18 Analog drum synthesizer.

The DRM1 MKIII is an amazing drum synth that is made in Erlbach, Germany. There are eight different sounds that are manipulated by a series of knobs on the front of the unit.

The back panel of the DRM1 MKIII has two audio outputs and a MIDI In/MIDI Through. There is an expansion pack that adds individual trigger inputs for each sound, but the MIDI is most important for sampling. There are also individual outputs for each channel on the front panel. The silver button on the left side of each sound is a trigger, which means you can manually trigger sounds without a MIDI input.

Sample robot is designed to create new instruments with very little effort. Its wizard walks you through all of the settings and then you simply start the process. Afterwards, the

Figure 2.19 Inputs and outputs.

Figure 2.20 Sample Robot Wizard.

Figure 2.21 Final instrument.

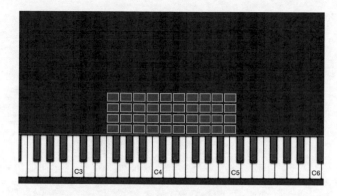

instrument can be exported to a file type that is compatible with the EXS24 for use in music production.

Nontraditional Sound Sources

One possible scenario when creating synthesizer-type instruments is to record nontraditional sound sources in the capture phase and then process them with traditional synth modules. The sources can be either everyday objects around your house or exotic, out-of-this-world-type sounds. Once you catch the sampling bug, you'll find yourself walking around listening to sounds and wondering if they would make good sound sources.

Examples

1. Metal Bowl

Metal bowls are capable of making very interesting sounds when spun on counters, scraped with utensils, and filled with various liquids. Experiment with different techniques to find a sound that is both interesting and has the potential to be a good synthesis source.

Equipment

A USB microphone is fed into Logic Pro X and different sounds are recorded. The most interesting sounds are edited and imported into the EXS24.

Figure 2.22 Kitchen bowl.

Figure 2.23 Screenflow capture sound samples online.

2. Comet Landing

The Philae lander made history by landing on a comet and, before stopping communication, it sent a recording of what it sounds like on the comet. This is a perfect example of a unique sound source that can be used as an instrument.

Equipment

The sound of the comet is available online as a video, so you won't need a microphone. Audio can either be looped from an available output on your computer to a different input, or you can use a screen capture program such as Screenflow to record the sound. In the example, we use Screenflow and then import the sound into Logic for editing and preparation for the sampler.

3. Vocalization

Perhaps the most flexible sound source is the human voice, which can be twisted and morphed into many different sounds. This example requires you to record sounds using your voice, and the editing and mapping forces you to decide if you want to make a percussive or melodic instrument and if you want to manipulate the recordings to fit across the range in Logic or use your voice to create them all.

Equipment

A USB microphone is perfect for this and it is often a good idea to use a pop-filter to prevent the vocalizations from puffing on the mic. The EXS24 is also used in this example.

Figure 2.24 Never discount your body as an excellent source for sampling.

Sample-Based Synthesis

Some digital synthesizers use sampled waveforms instead of oscillator models, which are perhaps the easiest way to create sound sources, but means that you won't have any elements of unpredictability. Instruments that do this are often less interesting, but in the case of a vintage instrument, you can "borrow" a little bit of its sound for a new instrument.

It can be hard to tell if a digital synth uses a recording of a single waveform of a different analog synth's oscillator unless it is published information, and even if a recording is used, it doesn't meant it is a good or bad instrument. Synth sources come in all shapes and sizes and a recorded waveform can be an excellent starting place for synthesis. It may be worth adding both analog and digital synths to your collection so that you can have a plethora of options in your sonic tool kit.

Historical Perspective on Sampling

In the quest to make music using technology, there has been a division between those who use these innovations to enhance popular musical production techniques and those that have sought to redefine what music is by breaking away from traditional conventions. The effort to reproduce the sounds of acoustic instruments in a workstation environment is one of the issues at the core of the division, but even though sampling has enjoyed widespread adoption, even newer options such as modeling are able to overcome its limitations.

At the time when electronic music was still in its infancy, inventors such as Harry Chamberlin and the Bradley brothers in the UK used analog tape to record individual notes of instruments and then created a mechanism attached to a piano-style keyboard to play them back. The most popular version of this sampler is called the Mellotron and it can be heard on a wide range of records starting in the era of the Beatles up to the present day. Accurate representations are now available in your favorite DAW and even on iOS for your iPad.

As digital technology evolved and improved, instruments such as the Fairlight CMI computer improved sampling, but remained expensive and impractical. Even though the original models had low resolution digital specs, it could reproduce incredible sounds that were used by artists such as Stevie Wonder and Peter Gabriel.[1] This line of sampling and music production tools has survived a rocky road and is still being developed. You can install and use an accurate replication on your iPad, with all of the quirks and limitations that came along with the popularity.

Computer-based sampling and hardware-based sampling have been in competition ever since, starting with the DX-1 from Decillionix that worked with both the Apple II in the early 1980s and the frontrunner hardware developer AKAI. Early AKAI samplers and then subsequent MPC workstations have set the standard for hardware, but software options have been able to leverage modern processing power to create sonically detailed instruments that are very realistic.

Gigasampler and Gigastudio from Nemesys, and later Tascam, proved that software sampling could work and provide full-length, high-quality samples for music production, but in a sad turn of events, it was discontinued after sample developers withdrew their support and moved towards Native Instrument's Kontakt sampler. The core issues were copyright protection and poor product management, leaving composers and producers to find other solutions.[2]

Samplers in the software realm have changed the music world forever, giving access to any instrument you can think of at your fingertips, with the ability to produce an entire orchestra on a single computer. It wasn't easy at first because of limitations in drive speeds, RAM amounts, and overall processing power. Every instrument had to be optimized very carefully to be playable, and so in the 1990s, a number of hardware developers used sampling technology to create sample playback units that allowed musicians to be able to have access to instruments without requiring a significant drain on computing processes. E-MU created a popular line of modules under the label Proteus, and Roland also had a long line of devices.

A MIDI studio would have a computer sequencer at the center and multiple external devices for sound creation all feeding into a mixing console. This type of studio could overcome the limitations of poor processing by compartmentalizing the roles of each part with hardware/software that matches in capability. As computing power increased, developers made consistent efforts to transition the sampling experience into software, and eventually external samplers and sample playback devices began to disappear. At first, you would have to use multiple computers to handle the large projects of orchestrators and composers, but the latest computers are able to handle almost everything by themselves.

Software samplers are now almost all universal in their ability to play back sample libraries in various formats. For example, the EXS24 in Logic Pro X can still import Giga libraries from the now discontinued Gigastudio. This makes the choice of sampler or DAW less important because each one can do

Figure 2.25 Gigastudio import option.

similar things; however, sample libraries are still highly individual and varied in what they offer.

For a period of time, musicians were likely worried that samples might reduce the amount of work they could get in the studio because a single MIDI programmer can play piano, bass, drums, and nearly everything else. While this is true and in fact a number of projects use sampled and synthesized sounds exclusively, samplers haven't been able to reproduce the subtle nuances of performance to match skilled musicians. Even in cases with detailed sample libraries that have the potential to sound as good as the real thing, it often requires more time and effort to do the same thing that a group of live musicians could do in a short amount of time.

So it is the list of limitations in sampling that defines its current place in music production because some say it has reached the peak of what it is able to do. Instruments can be sampled on every note, at many velocities layers, and with complex round-robin/key-switching implementation. A few samplers even offer advanced scripting features that mimic human performance. In spite of all of the advances over the years, sampled instruments have not completely replaced live, human performances. While new sample libraries are being consistently released, the future of virtual technology is modeling and nothing is likely to ever fully replace live musicians.

The one task that remains an important resource with sampling is creating patches of synthesizers for use on the stage and even in the studio. Instead of dragging every big synth with you on the road, you can sample the patches and play them on your laptop. Some functionality is lost because performing with a synth involves actively manipulating filters and effects, but these can also be emulated using effects in your software. An example of this is an iPad app created by Jordan Rudess of Dream Theater: it is a set of 57 instruments that he has used on albums and on tour. The app is called

Figure 2.26 Jordantron on iOS.

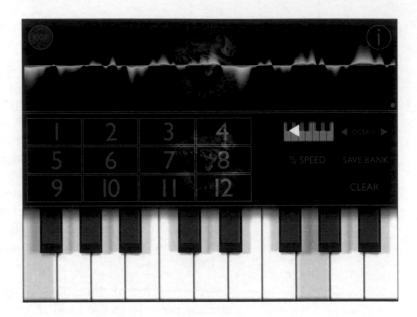

Jordantron and is an intimate look into the sounds that Jordan has sculpted over the years.

The Future?

Where is sampling going to go next? The basic technology hasn't changed in some time, and yet there is so much potential to improve what could be done. The promise of sampling has always been that the home studio enthusiast could create instruments without needing to know how to use computer programming languages. Modeling is the next step in evolution for the needs of musicians, but sampling still has value if only someone would invent a fully automated system for created instruments. The problem is that the money required for development would never be recouped because it isn't something that people are banging down the doors to buy.

This chapter is near the beginning of this book because sampling and synthesis have been intertwined for a long time and many of the same modulation and effects apply to both types of instruments. Perhaps this is an overgeneralization, but most synth geeks have also dabbled in sampling, and understanding both processes help form a more well-rounded musician.

Notes

1. "History (and future)," November 19, 2014. Retrieved from http://petervogelinstruments.com.au/history/
2. "Are Composers to Blame for GigaStudio's Demise? Some Observations," November 19, 2014. Retrieved from http://www.filmmusicmag.com/?p=1770

MODULATION
SYNTHESIS

3

Synthesizers that use modulation as their fundamental sound source are capable of creating exciting sounds; they remain unlike any other instrument since being introduced, and they continue to be a source of experimentation and sonic depth. Widely used in the 1980s and 1990s, Modulation Synthesis set the standard for digital synthesis and proved that synths could realistically replace instruments such as basses and pianos in pop arrangements.

Types

There are five types of modulation that this chapter explores, and in spite of a substantial amount of overlap, each provides a unique sonic palette. Modulation as a principle of synthesis has been a part of every major synthesis type as a secondary tool since the beginning of commercial synthesis. Along the way, designers realized they could give modulation a more prominent role in the tone-creation process, so a number of new synths were developed and released. Some efforts succeeded and have a sustained presence in modern practices, while others types of modulation have remained in a supporting role.

The most prominent of all synths that have modulation at their core is the DX-7 from Yamaha, which was released in the 1980s. It was labeled as a Frequency Modulation synth (FM) and was so popular that it single-handedly cemented both digital synthesis and MIDI as a part of the music production toolkit. The DX-7 is still a synth that is available and remains

Figure 3.1 The famous DX-7 from Yamaha.

in use, but there is still some controversy over which Modulation Synthesis that it uses.

The DX-7 is and always has been the most complex and tweakable preset keyboard ever invented. You have control over the entire sonic process, but most people only use the standard presets that shipped on the instrument. The default bass sound was used on hundreds of records and is arguably the most famous synth sound ever. Artists such as Queen, The Cure, Tina Turner, Michael Jackson, and so many more all used the DX-7 on their records and made this an incredibly popular and recognizable instrument.

The DX-7 has been called a FM synthesizer, but, if you look closer, you see that it actually uses Phases Modulation, which is related to the way it modulates the Operators that the DX-7 uses. If you Google FM Synthesis online, then you'll find a bunch of sites and books that present the math behind the sound, but this is outside the scope of this book and outside the scope of many peoples' ability to understand. Since the math doesn't help us understand how to craft specific sounds, we are not including it here. Phase Modulation and Frequency Modulation are nearly interchangeable, but FM Synthesis wouldn't be able to maintain the same fundamental frequencies in a multioperator system like Phase Modulation

Figure 3.2 DX-7 Algorithms.

can. But we are getting ahead of ourselves, so let's take a step back and explore the DX-7 in order to understand the full potential of FM and PM Synthesis.

The DX-7 uses six sine waves in various combinations to create all of the sounds it is capable of making. It doesn't have any filters or other effects, and it doesn't need them. The six sine waves, called Operators, are modulated together to create a wide range of sounds. There are 32 Algorithms that are used to determine how the Operators are combined. The relationship among Operators is defined by the roles of Carriers and modulators. The Carrier is the initial sine wave that is modulated by the modulator. The 32 Algorithms can have as few as one Carrier and five modulators, or as many as six Carriers and no modulators. It makes sense to think of each Carrier (and associated modulators) as individual synth engines that are combined together.

In Algorithms with two Carriers, it is possible to create layered sounds or patches with two segments. You could use the first Carrier/modulator to create the attack portion of a sound and the second Carrier/modulator for the body of the sound. With three Carriers, this concept expands, but the more Carriers you use, the simpler each individual element will be. If all six Operators are used as Carriers, then all you have are six

Figure 3.3 Data entry slider.

sine waves that are mixed together and the resulting sound has very little complexity.

In Chapter 1, LFOs were described as having the ability to modulate a target's pitch, amplitude, or filter cutoff. LFOs are generally low enough frequencies that they are inaudible, but their effect on other parameters can be substantial. The Operators in the DX-7 work in a similar way, where the modulator (which is in the audible frequency range) modulates the Carrier's phase. In audio, phase and frequency are intertwined, just as distance and time are connected with a moving object.

The DX-7 set the foundation for what remains the most thorough modulation experience available. The Algorithms that have minimal Carriers are loaded with modulators upon modulators. One modulator often modulates a second one before finally modulating the Carrier. A modulator might even have a feedback loop, which means it will be able to modulate itself. The level of complexity in some of the Algorithms is quite high and the resulting sounds are often quite exciting. The change in timbre resulting from the implementation of different Algorithms and modulator/Carrier ratios is often described mathematically through the use of a modulation index. This index is a perfect example of how complex the DX-7 is, both when programming patches and understanding why those patches sound like they do.

Figure 3.4 Modulators and Carriers.

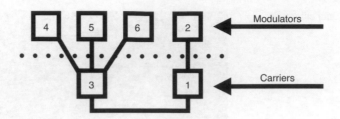

The tonal relationship between the Carriers and their modulators determine the complexity of the resulting sound. When audible frequencies are modulated together, the waves are inherently changed, and when certain waves are combined, then sideband frequencies are introduced into the resulting sound, which often matches fairly closely with bell-like sounds and metallic sounds. The last element of the DX-7 that is important to understand before we begin exploring how to create sounds is the inclusion of unique envelopes.

Each Operator has an individual envelope that controls its amplitude over time. Considering the complexity that an Algorithm may include, adding motion to the amplitude of each Operator has the ability to shake things up. The envelope generators (EG) are nontraditional and are both powerful but complex to operate. Each EG has four stages, each with Rate and Level settings. The Levels are set to any number between zero and 99, and then the Rates determine how long it takes to reach the next Level. There isn't a specific time correlation with the Rate settings because they change depending on the distance to the next level. $T = d/r$ is a formula that describes the time corresponding to the Rate setting, where T is time, d is distance, and r is rate. If the EG transitions from a Level of 60 to 70, it will happen quicker than from 40 to 80, even when the Rate is set identically.

A very exciting feature of this EG system is that you can create complex delays in Operator levels in a set pattern, which is useful for pseudo sequences. Using an Algorithm with all Carriers set at different pitches, each one can be set to varying

Figure 3.5 DX-7 envelopes.

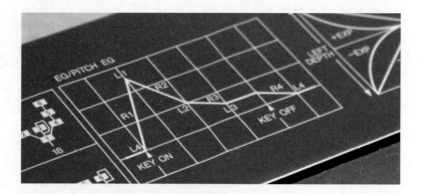

initial Rate settings, which turns on each of them at different times. This enhances the sound creation process with so much more than just FM Synthesis. On the DX-7, you should never set the Level below 50 because it will cause erratic behavior and a value of 50 is essentially an inaudible level.

The modulation process is flexible enough that very musical sounds can be created that mimic keyboards or basses, but it can also create screeching leads that sound like they come from another planet. Let's look at an example of how this process works using a DX-7 as an example.

Example 1: Algorithm 1/ Operators 1 & 2

Explanation

- Operator 1 (Carrier) is set to 1.0, which is the fundamental frequency.
- Operator 2 is shown with four different settings, 3.0–6.0, which are set as harmonics based on the fundamental of 1.0 (ratio setting).

Results

Each subsequent modulator increase results in a more complex sound, particularly in the higher frequencies. All four sound similar.

Figure 3.6 Harmonic content of various modulator combinations.

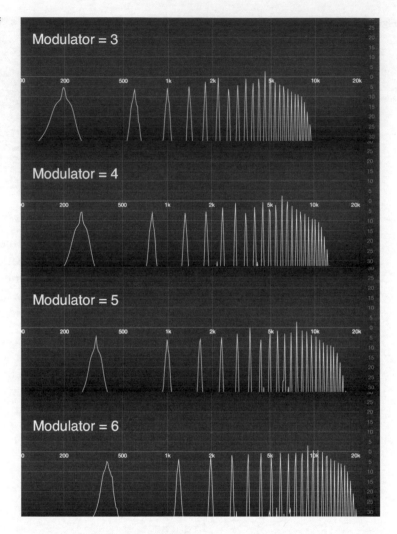

The DX-7 is a classic synth and is well suited to demonstrating the power of Modulation Synthesis. It isn't, however, going to be a synth that has patches, which translate very well to other FM/PM synths. Most synths that have FM capabilities only use a single Carrier/modulator, which is still capable of very typical sounds, but not to the extreme that a six-Operator system can. Newer software synths often add extensive features, which make it more of a hybrid synth rather than solely an

FM synth. FM8 from Native Instruments or Operator from Ableton both do amazing things with Modulation Synthesis, but they incorporate Subtractive Synthesis and other types of synthesis and they are examples of the power that soft synths are able to offer.

Once the capabilities of the DX-7 are explored, then we'll branch out into the other Modulation Synthesis offerings such as Amplitude Modulation (AM), Ring Modulation (RM), and Phase Distortion Modulation (PDM). Each of these is very similar to FM Synthesis, and, yet, they haven't remained prominently active as modern synth offerings.

Creating a DX-7 Patch

The first step in creating a new, DX-7 patch is deciding what you want it to sound like. That may seem obvious, but unlike Subtractive Synthesis, which lends itself better to tweaking and twiddling, the DX-7 is more like a black hole unless you approach it with a specific plan in mind. The three elements that you should plan for are sound segments, layers, and length.

Sound Segments

The number of desired segments of your sound will help determine the Algorithm that you choose. Here are several examples:

Hammond organ example

1st segment—percussive attack.
2nd segment—sustained note with layers.

Bass example

1st segment—sustained note body.

Piano example

1st segment—slight percussive attack.
2nd segment—sustained note.
3rd segment—decaying release.

Each example demonstrates the differences between sounds that require alternate Algorithms on the DX-7. Is it possible to create an accurate acoustic piano clone on the DX-7? No, but it is possible to have patches that resemble the real thing. Unfortunately, there isn't a clear path to creating those types of sounds without extensive trial and error.

Layers (and Texture)

After deciding on the appropriate number of segments, you'll need to pick the Algorithm that has the right composition of modulators. Each segment needs to achieve a specific sound quality, and only some Carrier/modulator combinations will work. If the middle segment needs a complex sound for a metallic pad, then choose an Algorithm that has at least one Carrier set with multiple modulators. The DX-7 displays the Algorithms with Carriers that are numbered, but those numbers don't necessarily determine the order of playback timing in the final patch; you can set those independently with the envelopes.

If the patch you envision requires additional layers, you can utilize another Carrier group that is set to a different frequency or texture, which is assigned to the exact same timing

Figure 3.7 DX-7 librarian.

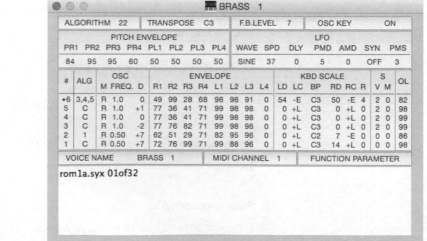

BRASS 1

ALGORITHM 22	TRANSPOSE C3	F.B.LEVEL 7	OSC KEY ON

PITCH ENVELOPE								LFO						
PR1	PR2	PR3	PR4	PL1	PL2	PL3	PL4	WAVE	SPD	DLY	PMD	AMD	SYN	PMS
84	95	95	60	50	50	50	50	SINE 37	0	5	0	OFF	3

#	ALG	OSC M FREQ. D			ENVELOPE R1 R2 R3 R4 L1 L2 L3 L4									KBD SCALE LD LC BP RD RC R							S V M		OL
•6	3,4,5	R 1.0	0	49	99	28	68	98	98	91	0	54	-E	C3	50	-E 4	2 0	82
5	C	R 1.0	+1	77	36	41	71	99	98	98	0	0	+L	C3	0	+L 0	2 0	98
4	C	R 1.0	0	77	36	41	71	99	98	98	0	0	+L	C3	0	+L 0	2 0	99
3	C	R 1.0	-2	77	76	82	71	99	98	98	0	0	+L	C3	0	+L 0	2 0	99
2	1	R 0.50	+7	62	51	29	71	82	95	96	0	0	+L	C2	7	-E 0	0 0	86
1	C	R 0.50	+7	72	76	99	71	99	88	96	0	0	+L	C3	14	+L 0	0 0	98

VOICE NAME BRASS 1	MIDI CHANNEL 1	FUNCTION PARAMETER

rom1a.syx 01of32

information. If you need two segments, but want to layer sounds, then you might find that a four-Carrier Algorithm is needed to make the sound; one would be used for the first segment and three for the layered second segment.

Length

You should have the Algorithm chosen before assigning envelope lengths to the individual Operators. The DX-7 uses a unique envelope system that has five rates with five associated levels. The first segment can start immediately, while the second segment is delayed at a zero level for a set amount of time. Using the envelopes to time each segment appropriately, you can ensure that each portion happens exactly when you want it to.

In Subtractive Synthesis, it is possible to create harmonic motion though the use of envelope-controlled filters or modulation with LFOs. The DX-7 has neither, so motion over time is created by assigning envelopes to the levels of Operators so that they shift modulation intensity as the note is sustained. The end result is much different than the movement of a filter's cutoff frequency, but the effect can be sculpted to play the same role of creating movement in the patch.

Miscellaneous Features

Pitch modulation on the DX-7 is limited to global parameters, which makes sense given that changing the pitch of individual Operators midstream would likely result in very unpredictable and unmusical sounds. Of course, that may be exactly what you want, and several software synths are able to do just that.

The DX-7 has a number of other performance aids, such as adjustable sensitivity for key velocity, keyboard level scaling, a pitch bend wheel, a mod wheel, and an input for breath control. None of these features are integral to the Modulation Synthesis process, but they are nice to have when using the DX-7 as a performance instrument.

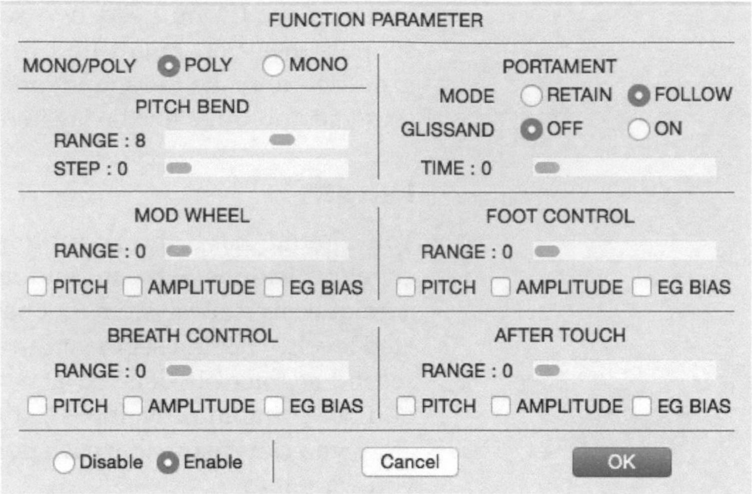

Figure 3.8 DX-7 parameters.

Programming the DX-7

The DX-7 is not fun to program because of the tiny screen and the single data entry slider. There are a lot of menus that you have to access and the entire process is very tedious. In addition to a difficult interface, tweaking parameters is an exercise in mental toughness in trying to remember which Operator is doing what, and keeping track of everything without any visual cues. There are blank patch sheets available online that you can print out and write on to help keep things organized, but there are also software control apps that are very helpful in the programming process. Either way, the more you use the DX-7, the better you'll be able to navigate its unique programming structure. After the DX-7, Yamaha released a number of different instruments that use FM synthesis, some of which have attempted to increase the programmability, but none of which ever became as popular as the original. Head to the recipe section for examples of FM programming.

Other Types of Modulation Synthesis

Amplitude Modulation and Ring Modulation are both used as tools on synthesizers to create interesting sound sources, but are never the defining element. The Arp 2600 has both a

Figure 3.9 Arp 2600 modulation.

Ring modulator module and the flexibility to use one of its three oscillators for AM, but the instrument is still considered a subtractive synthesizer.

Amplitude Modulation

Amplitude Modulation is the most common carryover from LFOs in Subtractive Synthesis, but it is also capable of creating very useful and interesting sounds. An LFO is used to create tremolo by slowly modulating an oscillator's level (or pitch for vibrato), but when you speed up the modulator into the range of audible frequencies, you begin to create brand new sounds. A key distinction is that AM synthesis doesn't use the same bipolar oscillation that you typically find in oscillators and LFOs, and instead uses a unipolar modulator that keeps the phase in the positive between 0–1. This difference isn't important when using a module designed for AM synthesis, but it helps explain the difference between AM and Ring Modulation, which is discussed later on.

The frequency of the modulator in AM synthesis functions very similarly to a two-Operator FM synth, but it's changing the amplitude level instead of frequency. There aren't any

Figure 3.10 AM synthesis example using the modulation matrix in Logic's ES2. LFO 1 is set to control Osc1Level.

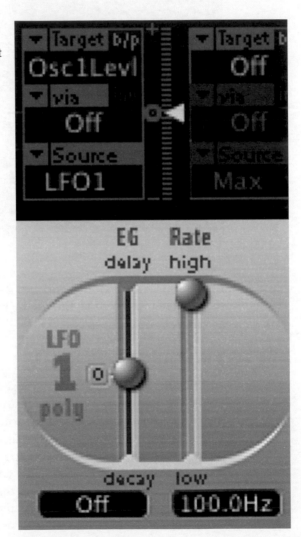

examples of 3+ Operator AM synth modules; in fact, there aren't any examples of synths at all that are labeled as AM synths. The only exception is with individual modules or as an additional feature of other synths.

AM Evolution

The modulation of the level of signals has been a widely used technique since the earliest days of synthesis, primarily in the form of LFO control. The level oscillates according to the speed of the LFO, creating a tremolo effect that is a standard offering on most instruments. When the rate of modulation is sped up into the audible range, the resulting sound is more complex and additional frequencies are introduced.

Modern instruments are almost universally capable of AM through the use of full-range LFOs. While they are still called Low Frequency Oscillators, most extend into the audible range and are capable of interesting AM Synthesis. Many software synths have LFOs with a variety of wave shapes; the result is sonically complex modulation. As is discussed later in the chapter, the most interesting uses of AM Synthesis are possible when the modulator is modulated.

Using a modulator to change the level of a Carrier is only truly synthesis when the rate is such that sidebands are

Figure 3.11 An LFO is used to modulate the level of an oscillator in the Modular App.

introduced and multiple frequencies meld together. In the analog realm, AM Synthesis is limited, just as it is with the other types of modulation, but with software instruments and DAWs, there are quite a few additional ways to use AM functionality. See later in the chapter for a closer look at using modulation matrixes and track automation for more complex AM.

Figure 3.12 Ring modulator.

Ring Modulation

The next form of Modulation Synthesis is known as Ring Modulation. Although not necessarily a synthesis type in itself, Ring Modulation utilizes Modulation Synthesis and is an integral aspect of synthesizers both vintage and modern. Sometimes referred to as balanced modulation, Ring Modulation works by multiplying two signals together—one signal typically being a simple waveform, such as a sine wave. Many people think Ring Modulation gets its name from the bell-like ringing sound that ring modulators produce. In actuality, ring modulators get their name because an analog ring modulator circuit features a series of diodes that are connected in a ring-like shape.

A ring modulator will feature two inputs, typically named "X" and "Y," for signals to be fed into. Like with other forms of Modulation Synthesis, one signal will act as a Carrier while the other will act as an incoming signal. Once the signals are combined and multiplied together, the ring modulator will output the sum and difference of the two signal's frequencies. One of the biggest differences between Ring Modulation and all of the other types of Modulation Synthesis is that it requires both signals to be present in order to work. If you suppress the Carrier signal so that you only hear the modulated output, then when there are breaks between notes, there

Figure 3.13 Ring Modulation circuit.

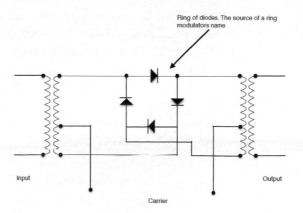

109

Figure 3.14 Ring Modulation diagram.

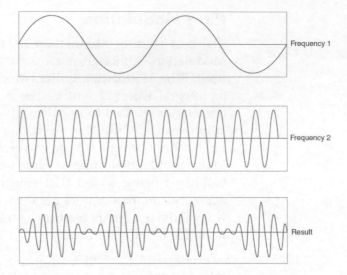

Frequency 1

Frequency 2

Result

won't be any extraneous sounds. When using a ring modulator, the resulting sound ideally doesn't contain the original source, but only a signal rich in harmonics created from the sum and difference of the two signals.

Ring Modulation in Synthesizers

Traditionally, ring modulators were used as an effect on synthesizers. Many analog subtractive synthesizers featured limited Ring Modulation capabilities in order to add extra tonal qualities to the overall sound. The Korg MS-20, for example, allowed users to combine the two oscillators using Ring Modulation in order to produce a slightly grittier sound. Many other early subtractive synths made by the likes of Moog and ARP featured similarly limited Ring Modulation in order to create bell-type tones. Because of the inclusion of limited functionality ring modulators, many synthesists felt that Ring Modulation was more of a gimmick than a powerful synthesis tool.

Despite the gimmicky vibe many people get from ring modulators, they are actually quite useful and powerful when used by someone willing to put in the time to experiment with their

Figure 3.15 MS-20 Ring Modulator.

When selecting "ring" on oscillator 2, the square wave of one oscillator is modulated via the square wave of the oscillator in a ring modulator circuit.

vast capabilities. Ring modulators, for example, are fantastic at creating bell and gong-type tones. Although FM synthesis most often comes to mind for creating bell-like tones, Ring Modulators will impart a much grittier and almost distorted aspect to the sound from the get-go. Because the inputted signals will not be present at the output stage of a ring modulator, extremely unique and weird sounds can be produced that are nearly impossible to re-create with any other synthesis technique.

A great example of Ring Modulation can be heard on the opening track of the movie *A Clockwork Orange*. Walter (now Wendy) Carlos used the Ring Modulator on her Moog Modular system in order to create percussive elements to the electronic rendition of "Music for the Funeral of Queen Mary" originally written by Henry Purcell. The first note on Carlos's electronic rendition is accompanied by a ring modulated, gong-like sound that can really only be appreciated when

Figure 3.16 Bode Ring Modulator.

heard. Carlos used her technique of ring modulated percussive sounds throughout her whole body of work. The score for the science fiction classic *Forbidden Planet*, created by Louis and Bebe Barron, also made great use of Ring Modulators.

Ring Modulators are actually quite old in design, but it was not until diodes were used in their circuitry that Ring Modulators were found to be useful in sound synthesis. One of the first synthesizers to use a Ring Modulator was the Melochord built in 1947 by Harald Bode. Due to the success of Bode's Ring Modulator, he went on to design a designated, stand-alone Ring Modulation known as the Bode Ring Modulator. The Bode Ring Modulator was later built in module format for inclusion in Moog modular systems. Buchla, Oberhiem, EMS, and Yamaha all followed suit and began including ring modulators into their synthesizers. Ring modulators are by no means only found in analog synthesizers. Ring modulators can be found on virtually every synthesizer. Ring modulators have become a synthesis staple and, for that reason, they are found on a wealth of new synthesizers both analog and digital based.

Phase Distortion Modulation

The DX-7 was extremely popular, but its technology was protected by a patent, and so other companies couldn't release FM synths that used FM/PM without getting in trouble. Casio came up with a related synthesis process called Phase

Figure 3.17 Casio PDM.

Distortion Modulation, which meant they could achieve a similar sound without it infringing on the patent.

Casio is the only developer to explore PDM and, while it is capable of creating sounds which sound just like FM and PM offerings, it expanded on the basic premise of existing modulation techniques by adding more complex waveforms. The DX-7 relied on a patent that identified the Carriers and modulators as sine waves, and Casio expanded into other waveforms such as the triangle and the square. These modulate the phase of additional waveforms by adjusting their phase playback. A change in the modulator results in a faster or slower playback of the Carrier, which creates complex distortion.

PDM is similar to another type of synthesis, which is covered in a different chapter, called Wavetable synthesis in that the primary waveforms are not played back in linear fashion, but the timing is changed through modulation. The changes in time occur in a cyclic manner, which is the reason that it resembles FM synthesis. Casio's PDM is particularly similar to the DX-7 algorithms, which have multiple modulators in a single chain because, in both instances, the Carrier is modulated by complex waveforms; PDM uses sounds that are more complex at the source, and the DX-7 builds complex modulators by having sine waves modulate each other. The end results are not identical, but can be tweaked to be very similar.

The basic waveforms used in PDM are as follows:

Saw-tooth
Saw-tooth alternates
Square
Pulse
Double Sine
Saw-tooth and pulse combo
Triangle
Trapezoid

Systems typically have only two oscillators, but that is enough to create a wide range of sounds with the available tones.

Modulation in DAWs

Software innovation has increased the power of Modulation Synthesis by providing more ways to connect various components and certainly more flexibility while doing so. Most digital synths, such as the DX-7, are limited in the modulation process by their design, which means you can't use an external option to modulate the Operators in the Algorithms. Software design removes the limitations of the hardware instruments and expands through the incorporation of modern automation.

Automation is attached to user-determined destinations and is easily assigned to control instrument parameters. The sync consideration that is important is whether you want the modulation to speed up or slow down depending on the triggered pitches, or if you want a stationary modulation rate. Using an automation track, you can set a modulation rate, but it is

Figure 3.18 Automation example—controlling amplitude via automation curve.

more difficult to set it to vary with the pitch. The instrument itself is more likely to have a KBD assigned modulation variance that follows the pitch of the triggered notes.

Modulating the modulators is one way of using MIDI control and automation in your DAW. Incorporating the tools of the DAW into the synthesis process means creating an experience that is semimodular, much like patching in a new module on an old Moog. Many DAWs also have a mapping feature that connects physical controllers with software destinations, and then the modulation parameters are controlled by external means.

The key is to keep track of all modulation sources and destinations and then to realize when too much is too much. You can create some very intense tones, but without having a specific reason to do so, you might end up with an unusable sound. The key to crafting the sound you want is to have the end sound in mind, to understand the tools you have, and then to put forth the effort to achieve your goal.

Brief History of FM Synthesis

The history of modular synthesis is pretty nearly the history of synthesis and so it isn't practical to describe the evolution of each technique here. The context of FM Synthesis as it pertains to the DX-7 is worth exploring because it made an impact on popular music and beyond. John Chowning studied and subsequently worked at Stanford University where he stumbled on FM Synthesis. John was a pioneer of computer music and studied the correlation between acoustic instruments and digital synthesis. One day, he was experimenting with vibrato at higher speeds and recognized the resulting tonal complexities. John himself has called the discovery naive, but his ability to recognize the importance of this discovery and then be able to describe it in a meaningful way is the contribution that is the most important.

FM Synthesis in the digital realm was put under patent and Yamaha recognized its value and put this new technology into the DX-7, which became one of the most popular synthesizers ever. The default sounds on the DX-7 were carefully crafted to sound like real instruments, which were latched onto by performers as a stage and studio instrument that replaced their other acoustic and expensive instruments. FM Synthesis may have started as a technique capable of complex synthesis, but it quickly became a tool that was used as preset piano and not as a synthesizer at all. Several of the default sounds have been used so often that they are easily recognizable and a staple of pop/jazz music in the 1980s and 1990s. You can hear the DX-7 on songs from Michael Jackson, Tina Turner, Brian Eno, and so many more.

The DX-7 has a horrible interface, which discourages musicians from programming new sounds, and, even though it was extremely popular, it has never been a bridge for musicians to enter the realm of synthesis. Once other keyboards came to market that could more realistically reproduce other musical instruments, the DX-7 lost its appeal and fell out of favor. The good news is that you can now buy an original DX-7 for a fraction of the cost of the original price and it is still an excellent instrument capable of creating wonderful sounds. You can even use a software editor to aid in the programming process, which is critical when experimenting with this instrument.

The DX-7 created a legacy of success that has often been sought after by other designers, but never replicated. Now that rights to FM Synthesis are open, there are a number of software instruments that accurately emulate the DX-7 and others that greatly expand on its toolset. Several software instruments can even exchange patches with the DX-7, and that makes programming it substantially easier.

The greatest success of FM Synthesis and Modulation Synthesis is that they provided us with complex results without requiring unreasonable amounts of computing power. To create the same complexity of sound with analog or additive synthesis would have been cost prohibitive in the 1980s and the DX-7 demonstrated a way to craft useable sounds out of minimal resources.

Recipes

The DX-7 is an entirely different beast than subtractive synths and it makes creating a common recipe format more interesting. The emphasis is on the Algorithm and the individual Operators, but with a substantial amount of information associated with each Operator, it makes it difficult to create a visual recipe that translates to all FM synthesizers. As with all of the included recipes, the primary objective is to help readers master the basics of synthesis: this section is no different. The exception is that an analysis of several default patches from the DX-7 are included to demonstrate basic techniques while also showcasing patches that have been used on many records and remain iconic in their sound.

Recipe 1: Brass 1

The initial patch on the DX-7 is the Brass 1 patch, which shows just how flexible the FM engine is in the creation of sounds, both realistic and not. In the case of Brass 1, the DX-7 takes a more traditional approach with a dash of FM synthesis mixed in. The core sound is a three-Carrier set all modulated by the same modulator. Each has the same basic frequency relationship, but all are detuned slightly, which gives the brass sound additional texture and depth. The modulator/Carrier

Figure 3.19 DX-7 Brass 1 settings shown in DX7 Librarian software.

ALGORITHM 22				TRANSPOSE C3				F.B.LEVEL 7			OSC KEY			ON
PITCH ENVELOPE								LFO						
PR1	PR2	PR3	PR4	PL1	PL2	PL3	PL4	WAVE	SPD	DLY	PMD	AMD	SYN	PMS
84	95	95	60	50	50	50	50	SINE	37	0	5	0	OFF	3

#	ALG	OSC			ENVELOPE								KBD SCALE						S		OL
		M	FREQ.	D	R1	R2	R3	R4	L1	L2	L3	L4	LD	LC	BP	RD	RC	R	V	M	
•6	3,4,5	R	1.0	0	49	99	28	68	98	98	91	0	54	-E	C3	50	-E	4	2	0	82
5	C	R	1.0	+1	77	36	41	71	99	98	98	0	0	+L	C3	0	+L	0	2	0	08
4	C	R	1.0	0	77	36	41	71	99	98	98	0	0	+L	C3	0	+L	0	2	0	99
3	C	R	1.0	-2	77	76	82	71	99	98	98	0	0	+L	C3	0	+L	0	2	0	99
2	1	R	0.50	+7	62	51	29	71	82	95	96	0	0	+L	C2	7	-E	0	0	0	86
1	C	R	0.50	+7	72	76	99	71	99	88	96	0	0	+L	C3	14	+L	0	0	0	98

VOICE NAME	BRASS 1	MIDI CHANNEL 1	FUNCTION PARAMETER

relationship is simple, which is common in sounds that are designed to emulate acoustic instruments and are musical in nature.

Recipe 2: Bass 1

This is the most famous patch on the DX-7 and has been used on countless records over the past few decades. It doesn't get used as much as it once did, but it's hard to argue with artists such as Michael Jackson, Madonna, Depeche Mode, and so many more. The parameters that are involved with making this patch are not as intuitive as you might guess, and the recipe format had to be expanded to cover all of the elements.

Figure 3.20 DX-7 Bass 1 settings shown in DX7 Librarian software.

ALGORITHM 16			TRANSPOSE C2				F.B.LEVEL 7				OSC KEY			ON
	PITCH ENVELOPE										LFO			
PR1	PR2	PR3	PR4	PL1	PL2	PL3	PL4	WAVE	SPD	DLY	PMD	AMD	SYN	PMS
94	67	95	60	50	50	50	50	TRI	35	0	0	0	OFF	3

#	ALG	OSC			ENVELOPE								KBD SCALE					S		OL
		M	FREQ.	D	R1	R2	R3	R4	L1	L2	L3	L4	LD	LC	BP	RD	RC	R	V	M
•6	5	R	9.0	0	94	56	24	55	93	28	0	0	0	-L	A-1	0	-L	1	7	0	85
5	1	R	0.50	0	99	0	0	0	99	0	0	0	75	-L	C#4	0	-L	7	3	0	62
4	3	R	5.0	0	90	42	7	55	90	30	0	0	0	-L	A-1	0	-L	5	5	0	93
3	1	R	0.50	0	88	96	32	30	79	65	0	0	0	-L	A-1	0	-L	6	3	0	99
2	1	R	0.50	0	99	20	0	0	99	0	0	0	0	-L	D3	0	-L	7	0	0	80
1	C	R	0.50	0	95	62	17	58	99	95	32	0	57	+L	A2	14	-L	7	0	0	99

VOICE NAME	BASS 1	MIDI CHANNEL	1	FUNCTION PARAMETER

Figure 3.21 DX-7 E. Piano settings shown in DX7 Librarian software.

ALGORITHM 5			TRANSPOSE C3				F.B.LEVEL 6				OSC KEY			OFF
	PITCH ENVELOPE										LFO			
PR1	PR2	PR3	PR4	PL1	PL2	PL3	PL4	WAVE	SPD	DLY	PMD	AMD	SYN	PMS
94	67	95	60	50	50	50	50	SINE	34	33	0	0	OFF	3

#	ALG	OSC			ENVELOPE								KBD SCALE					S		OL
		M	FREQ.	D	R1	R2	R3	R4	L1	L2	L3	L4	LD	LC	BP	RD	RC	R	V	M
•6	5	R	1.0	+7	95	29	20	50	99	95	0	0	0	-L	D3	19	-L	3	6	0	79
5	C	R	1.0	-7	95	20	20	50	99	95	0	0	0	-L	A-1	0	-L	3	0	0	99
4	3	R	1.0	0	95	29	20	50	99	95	0	0	0	-L	A-1	0	-L	3	6	0	89
3	C	R	1.0	0	95	20	20	50	99	95	0	0	0	-L	A-1	0	-L	3	2	0	99
2	1	R	14.0	0	95	50	35	78	99	75	0	0	0	-L	A-1	0	-L	3	7	0	58
1	C	R	1.0	+3	96	25	25	67	99	75	0	0	0	-L	A-1	0	-L	3	2	0	99

VOICE NAME	E.PIANO 1	MIDI CHANNEL	1	FUNCTION PARAMETER

Recipe 3: E. Piano

This is another classic DX-7 patch that has been heard on hundreds of records in the 1980s, including artists such as Chicago, Phil Collins, and Luther Vandross. As with so many of the popular presets, this is another attempt at re-creating another instrument, but with a DX-7 flavor.

ADDITIVE SYNTHESIS   4

Additive synthesis can be considered the holy grail of synthesizer formats. Additive supplies the user with the greatest amount of options and sonic possibilities when creating sound. In theory, any sound imaginable is able to be re-created with the utmost precision and accuracy when using an additive synthesizer. No other synthesis formats come close to its gargantuan nature. That being said, additive synthesis is one of the more complex synthesizer formats and therefore requires a deeper understanding and sharper learning curve in order to master it. Once mastered however, the possibilities are literally endless in sound creation and re-creation. In simple terms, additive synthesis is the ultimate sound-creation tool.

Additive Synthesis Theory

As briefly mentioned in the subtractive synthesis chapter, any sound, be it a trumpet, a voice, or an explosion, is made up of a series of harmonics. In essence, these harmonics are all individual sine waves that are multiples of the fundamental frequency of the sound. These harmonics differ in amplitude and also change amplitude over time, resulting in the complex sound that we hear. Additive synthesis utilizes this by allowing the users to combine a set amount of sine waves, as well as designate their amplitudes, in order to create sound.

Additive synthesis has its roots in the Fourier series. Jean-Baptiste Joseph Fourier is credited with conducting research that led to and inspired the Fourier series. A Fourier series is basically a way in which to represent a waveform as a combination of its various, individual harmonics in their most simple state—sine waves. Any wave analyzed via the Fourier series is subsequently broken down to its key elements. In practice, the Fourier series is often used as a way of displaying the various or even infinite independent sine waves that make up a sound. For this reason, additive synthesis is sometimes referred to as Fourier synthesis.

The Fourier series is often used in FFT (Fast Fourier Transform) units as well as a vast array of software devices, such as phase vocoders and resynthesis engines. Engineers theorized that if sound was made up of individual sine waves, each at different frequencies and amplitude as the Fourier series determines, any sound could hypothetically be resynthesized simply by adding in these same sine waves to their exact proportions. The theory behind additive synthesis is actually quite old and even predates subtractive synthesis.

By using additive synthesis, the user is free to adjust each harmonic and its amplitude, resulting in an extreme level of control. An additive synthesizer has a number of sine wave generators that are able to be set to any frequency or amplitude. These sine wave generators are used in order to create the various harmonics needed to build a sound. Early additive synthesizers were capable of producing only a few harmonics, while the first commercial additive synthesizers could produce up to around 64 individual harmonics. The advent of software additive synthesis has brought that number up to the many hundreds. In additive synthesis, the more harmonics you are able to produce, the better.

When building a sound, the user will tune each sine wave generator to correspond to various harmonics. By allowing the user to control every single harmonic in a sound, not only are incredibly accurate sound re-creations possible, but extremely complex and new sounds seem to seep from an additive synthesizer. Because sound is constantly changing and rarely remains stagnant, various types of amplitude modulation are employed on an additive synthesizer in order to bring life to a patch. In fact, if one were to simply set the amplitudes of the individual harmonics and leave them, the resulting sound would be stagnant and be reminiscent of an organ. For this reason, many additive synthesizers will feature independent envelope generators and LFOs for each harmonic, or set of harmonics, so the user can create extremely complex patches. By incorporating envelope generators and LFOs, the user can designate specific harmonics to rise or fall in amplitude before, during, or after the fundamental frequency is heard. By manipulating harmonic amplitude over time, incredibly complex filter sweep-type sounds can be produced that cannot be replicated on a subtractive synthesizer.

Chances are that many of you have used some form of additive synthesis without knowing it. In fact, just combing waveforms in the mixer section of a subtractive synthesizer can be thought of as a very basic form of additive synthesis. When combining various waveforms, say a sawtooth wave and a square wave, you are effectively creating a new complex wave via additive synthesis. In addition to this, many vintage analog subtractive synthesizers were capable of performing a small amount of additive synthesis. The Roland SH-3A, for

Figure 4.1 Creating filter responses through additive synthesis.

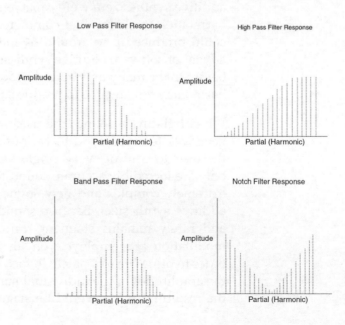

example, allowed the user to combine various octaves and wave shapes on its single oscillator as a means to create more complex waves. The Oxford Synthesizer Company's Oscar synth featured a more advanced form of additive synthesis. When using the synth in "harmonic creation mode," the upper register of the keyboard allows users to manually add in up to 24 harmonics. The user was able to hear the additional harmonics added in real time as the sound became richer. The newer Arturia MiniBrute analog subtractive synthesizer features similar additive capabilities to the Roland SH-3A in that the user can combine various wave shapes on the single oscillator. Although these basic forms of additive synthesis pale in comparison to what stand-alone additive synthesizers can do, they feature a form of additive synthesis none the less.

Additive synthesis is most often employed on digital or software-based instruments. By utilizing digital signal processing, a degree of precision is inherent that would not be available if one were to use analog circuitry. In an analog synthesizer, for example, temperature fluctuations can lead to pitch drifting in an oscillator. Although pitch drift is acceptable or even sometimes desired on a subtractive synthesizer, it can wreak havoc on an additive synth. Because additive synthesis incorporates so many individual sine waves in order to designate the timbre of the overall sound, the slightest unwanted pitch variation in any of the harmonics can cause dramatic changes to the sound as a whole. Digital signal processing allows for extreme pitch stability resulting in much more stable timbres.

Figure 4.2 The Oxford Synthesizer Company Oscar synth featured limited amounts of additive synthesis. Photo courtesy of www.perfectcircuitaudio.com.

Problems with Analog Additive Synthesis

In order to understand the inherent difficulties in having an analog additive synthesizer, let's go through what is needed in order to build an analog additive synthesizer. The first, most important thing needed for an additive synthesizer is a tone generation source. In the analog world, tone production, even sine wave production, is easy enough via an oscillator. For an additive synth, we would need a number of individual oscillators: the more, the better. Let's imagine a synthesizer capable of producing 20 harmonics—which is much greater than the earliest analog incarnations of additive synthesizers, but when compared to digital additive synthesizers that contain anywhere from 64 individual sine wave generators up to a few hundred, 20 is a rather conservative figure. In order to produce 20 individual harmonics, our synthesizer would need 20 individual analog oscillators. When one takes into consideration that the famous MiniMoog synthesizer only has three oscillators, it quickly becomes apparent how big a 20-oscillator system would be. Tone generation however, is only the beginning. The next thing we would need in our synthesizer is amplifiers, 20 of them to be exact. In order to be able to control each harmonic individually, a dedicated amplifier for each oscillator will be required.

Now we have a synthesizer able to create up to 20 individual harmonics, each with an individual amplitude. As was discussed earlier, just setting the amplitude of individual harmonics and leaving them results in stagnant, often boring sounds. So now, in order to add amplitude control over time to our individual harmonics, we will need 20, separate envelope generators. As great as envelope generator control is however, more means of modulation are necessary to create the interesting sounds that additive synthesis is capable of. In order to add another dimension of control to our synth, we will have to add 20 individual LFOs to our system so we can really manipulate each harmonic amplitude in a suitable way.

Figure 4.3 Our analog
20-oscillator system.

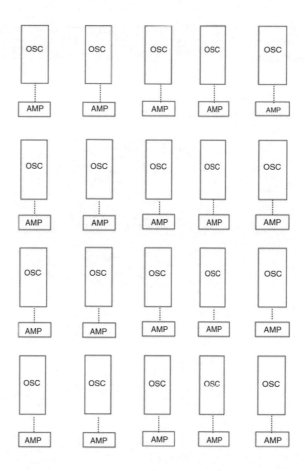

So now we're at a synthesizer capable of creating 20 harmonics, each with envelope and LFO manipulation. So far, our synthesizer contains 80 individual modules—that is no small number not even taking into consideration the costs. Despite this large number of modules, we are still not done building our additive synthesizer. All natural sound contains some sort of noise, such as the user's breath when using a woodwind instrument. Although noise is certainly able to be broken down to its individual sine waves using the Fourier series, the sheer amount of sine wave generators necessary to re-create noise would be unattainable. In order to add noise to our system, we will add a few noise generators capable of producing different noise shapes.

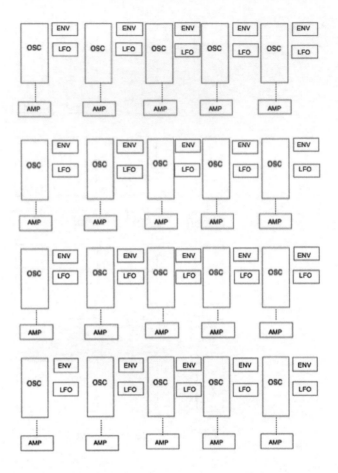

Figure 4.4 Our analog 20-oscillator system with 20 envelope generators and LFOs.

Finally, we will have to add some sort of summing bus to mix all our 20 harmonics together into a single output as well as a means to route our keyboard's pitch and gate information to the various oscillators and envelope generators. Although not exactly necessary, it might be desirable to add one final envelope generator and LFO in order to control the entire sound as a whole.

So now we have a respectable, 20-harmonic additive synthesis with all the bells and whistles one would expect. The sheer size, weight, power consumption, and cost are extreme, but we are free to make the sounds of our dreams. We are forgetting one thing though, because our synthesizer uses analog

Figure 4.5 Our analog 20 partial additive synthesizer.

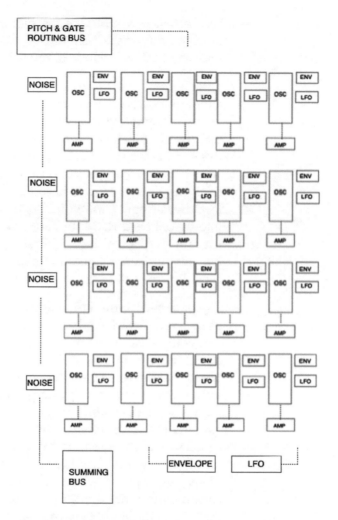

circuitry, every single circuit is susceptible to temperature and pressure changes, movement, power inconsistencies and a variety of other maladies. All these little problems that become amplified through these circuits cause our oscillators to drift in and out of tune as well as wreak havoc on our envelope generators, amplifiers, and LFOs. In all likelihood, our additive synthesizer we just built won't be very useful for re-creating sounds with exact precision or maintaining a sound we create for very long. It is quite apparent that additive synthesis and analog circuitry were simply not meant for each other. It would take the digital synthesis revolution to bring additive synthesis out of the universities and laboratories and into the hands of musicians and sound designers.

Early digital synthesizers that included additive synthesis, although powerful, were not very practical. Instruments such as the Fairlight CMI had advanced computational powers for the time, but lacked the ability to make changes to an additive synthesis sound in real time. The problem with not being able to make changes in real time was that the synth effectively acted as a sample-based synthesizer when played. In fact, the Fairlight CMI would load newly synthesized sounds into its onboard RAM, or random access memory, once the user pressed "compute." Therefore, these early digital additive synthesizers were oftentimes less than useful when one considers the full potential of an additive synth.

The first, successful digital additive synthesizer that allowed users a number of available harmonics with real time control and computing while remaining somewhat affordable came out of Japan—the Kawai K5. The Kawai K5 was revolutionary in that it offered 64 independent harmonics with a whopping eight notes of polyphony![1] Eight notes of polyphony might seem laughable today, but at its inception, only the Yamaha DX series offered more. When using the K5, a user could sacrifice polyphony by half in order to have 128 individual harmonics. Having this many harmonics was completely unheard of at the time, especially when taking into consideration that the K5 was somewhat affordable. In fact, the K5 and its 64 or 128 harmonics still stands today as a quite respectable additive synthesizer.

Arguably, one of the biggest drawbacks to additive synthesis is the amount of time it takes to program a sound. Each harmonic needs to be tuned and set at a specific amplitude and then amplitude modulation via envelope generators and LFOs need to be set. When working with an instrument that is capable of producing 64 or 128 individual harmonics, this is no easy or quick task. The Kawai K5 had some fairly unique

Figure 4.6 Kawai K5. Photo courtesy of www.perfectcircuit audio.com.

tricks in order to speed up the sound-creation process. Firstly, the K5 displayed each harmonic with a bar representing its amplitude (similar in look to a bar graph). By showing all of the harmonics and their individual levels visually, the user was able to quickly visualize what was happening with the sound. Next, the K5 allowed the user to group harmonics together in order to adjust them simultaneously. The K5 allowed users to select groups consisting of even harmonics, odd harmonics, octaves, fifth intervals, or custom groupings. The user was also free to manually adjust individual harmonics so that no customization was hindered by creating these groups—it just allowed for quicker adjustments. Allowing users to group various harmonics together was a completely new feature in the additive synthesis world. Once a bank of harmonics was selected, they could be raised or lowered in amplitude while maintaining their proportions with one another in order to maintain their harmonic hierarchy. The K5 was the first additive synthesizer to utilize this type of control.[2]

The Kawai K5 offered a wide variety of envelope generators in order to further shape the resulting synth patch. Like with the individual harmonics, the K5's envelope generators were easy to adjust as well. The K5 featured envelope generator short cuts for effectively copying and pasting envelope generator responses to higher harmonics so that the user didn't have to start from scratch every time.[3] The Kawai line of additive synthesizers has gained monumental status in the additive world similarly to the Moog and Arp synths of the subtractive world. Kawai has remained a leader in additive synthesis. Kawai updated their famous K5 synthesizer in 1996 with the K5000 line, which modernized and reintroduced the world to hardware additive synthesizers. The K5000 and subsequent models not only offered the users more harmonics, but higher levels of controls than the K5. Firstly, the K5000 worked by having multiple "sources" that can be assigned and played. Each "source" could be an additive synthesis engine or a sample playback engine. When used as additive synthesis engines, each "source" can contain up to 64 individual harmonics. When combining multiple additive synthesis "sources," a large number of harmonics are able to be combined. Next, each harmonic was paired with an individual amplitude envelope generator, allowing for much more control than was offered on the K5. Finally, virtually every additive synthesis parameter offered on the K5000 was assignable to various performance control features. For example, key velocity, or the pressure applied to keys, key tracking, or playing up and down the keyboard, as well as onboard knobs, sliders, and wheels can all be set to control various additive synthesis parameters. The Kawai K5000, although perhaps not as successful as its older sibling, helped bring additive synthesis back into the mainstream. The K5 and K5000's innovations have influenced virtually all modern software additive synthesizers.

Figure 4.7 Kawai K5000. Photo courtesy of www.perfectcircuit audio.com.

Types of Sounds Possible with Additive Synthesis

Describing the types of sounds that additive synthesis excels at is quite difficult. In theory, any sound imaginable is able to be created on an additive synthesizer as long as you have a sufficient number of sine wave generators, envelopes, and LFOs, as well as a deep understanding of additive synthesis. For this reason, it is nearly impossible to say which types of sounds additive synthesis are best at creating. That being said, certain types of sound are more often created on additive synthesizers than other synthesis formats. The first, and most common, sound type additive synthesis is used for is sound resynthesis. Because additive synthesis is capable of rebuilding any sound using the Fourier series, it is frequently used in this way. Although modern sampling synthesizers have oftentimes made sound re-creation obsolete using additive synthesis, the degree to which sound can be re-created accurately, as well as the amount of customization one can add to the re-created sound, is unique to additive synthesis. Therefore, many people still prefer using additive synthesis to re-create sounds rather than using a sample-based synthesizer. Rich, evolving pads are also frequently produced on additive synthesizers. Sonic soundscapes and complex, evolving drones reminiscent of Vangelis's *Blade Runner* score are also among the most popular patches designed on additive synthesizers. By controlling the envelope for each harmonic, pads are able to evolve in timbre in a way that is mostly unachievable using other synthesis formats that only have envelope control over amplitude or filter responses. When using an additive synthesizer, the user can create a pad that starts out dark and warm and then slowly evolves to feature shimmering high-end harmonics while the lower harmonics slowly begin to become frequency modulated and darker as midrange frequencies become prominent, all through the means of envelope control over individual harmonics. The possibilities are literally endless for creating evolving pads when using an additive synthesizer.

Additive synthesis can also theoretically be used in order to emulate classic analog synthesizers. Oftentimes, digital synthesizers, which utilize virtual analog emulation, can impart digital artifacts such as aliasing and stepping onto the sound itself, causing these synthesizers to not sound accurate. Many engineers have toyed with utilizing additive synthesis in order to re-create various analog sounds. Although using additive synthesis in this way has rarely been used on commercial instruments, many studies have been conducted that have shown that additive synthesis can in fact be used to emulate subtractive synthesis sounds.[4] It is possible that as the analog subtractive resurgence grows, software synthesizer companies may begin to implement additive synthesis into their instruments in order to create more accurate-sounding analog emulations.

Additive Synthesis in Practice

So now that we have covered a brief outline of additive synthesis, let's go more in depth on additive synthesis theory as well as all of the parameters one might expect to find in an additive synthesizer in an attempt to learn how to use it successfully.

Tone Generation

As stated earlier, additive synthesis uses a number of sine wave generators in order to build sound. Many instruments will call these sine waves either harmonics or partials. Some users might be asking themselves why additive synthesis is limited to sine waves. Why not have sawtooth, square, and triangle wave generators in addition to sine wave generators in order to create more complex tones? As we discussed, all sound can be broken down using the Fourier series. The Fourier series breaks sounds down into their most basic elements: sine waves, or harmonics. Therefore, if the additive synthesist wanted to have a screaming sawtooth lead sound, all he or she would have to do is build a few sawtooth waves. Because a sawtooth wave contains both even and odd harmonics, all the user would have to do is add in both even and odd harmonics to his or her sound, as well as set the amplitude of each harmonic in correspondence to the inversely proportionate relationship they hold with the fundamental frequency and presto—a sawtooth wave is created.

Although additive synthesis is capable of so much more than the basic wave shapes, it is a helpful exercise to build the basic wave shapes using an additive synthesizer. So let's go through each of the common wave shapes and deconstruct them to their basic elements and then build them back up using additive synthesis.

Triangle Waves

Because a triangle wave only contains odd harmonics (third, fifth, seventh, etc.), we would have to add in harmonic sine waves at each of these intervals. Thankfully, most modern additive synthesizers do the math for us and display harmonics via their harmonic number rather than making a user calculate what frequency each harmonic would be. However, if using an older model or making an analog modular additive synthesizer, one can figure out the harmonic number's frequency simply by multiplying the fundamental frequency

Figure 4.8 Harmonic content of a triangle wave.

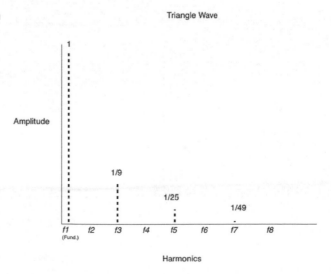

by the harmonic number. For example, if taking "A" 440, one would multiply 440 by the desired harmonic number (let's say the third harmonic). We would then multiply 440 by three and get 1320Hz, which is "A" 440's third harmonic. It must be remembered that a triangle wave's upper harmonics are proportional to the inverse square of the fundamental frequency; meaning, the third harmonic is 1/9th the amplitude of the fundamental while the fifth harmonic is 1/25th the amplitude. Using this inverse square proportion, we must lower the amplitude of each subsequent harmonic accordingly while also inverting the phase of every other sounding harmonic (i.e. third, seventh, eleventh). If this inverse square proportion and phase inversion is followed exactly, a perfect triangle wave will result. But additive synthesis is magical in the fact that we do not have to be confined to re-creating perfect wave shapes. In fact, one could be free to add in a few extra harmonics at a higher amplitude, resulting in something that is in between a triangle wave and a square wave. The possibilities are literally endless.

Sawtooth Waves

As stated above, a sawtooth wave contains both even and odd harmonics. Unlike a triangle, however, the upper harmonics

Figure 4.9 Harmonic content of a sawtooth wave.

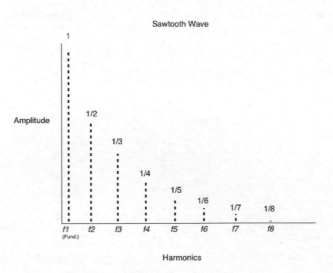

of a sawtooth wave are just inversely proportional to the fundamental frequency. This means that the second harmonic is 1/2 the amplitude of the fundamental while the third and fourth harmonic are 1/3 and 1/4 the amplitude respectively. When programming a simple sawtooth wave on an additive synthesizer, the user is free to follow these guidelines exactly, but again, due to the extreme customizable nature of additive synthesis, the user is free to experiment and make the second harmonic louder than the fundamental and each harmonic thereafter louder still. An interesting experiment to perform is to raise and lower the amplitudes of the upper harmonics once a sawtooth wave is created. The resultant sound will be reminiscent of a filter being opened and closed. This happens because all a filter does is effectively lower the amplitude of upper harmonics. Many classic analog subtractive synthesizers can be emulated this way. Emulating a filter response is only one of an endless amount of things one can do with an additive synthesizer.

Square Waves

Like a triangle wave, a square wave contains only odd harmonics. In order to create a square wave on an additive synthesizer, one must add in each odd harmonic. The rate

Figure 4.10 Harmonic content of a square wave.

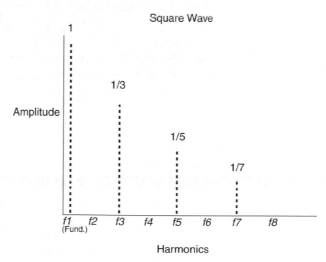

135

at which these upper harmonics diminish in amplitude, though, is much less than with a triangle wave. A square wave's upper harmonics will drop in amplitude at a rate that is inversely proportionate to the fundamental frequency or, in other words, the same as with the sawtooth wave mentioned above. Again, the additive synthesist is free to adjust these harmonics however they see fit, but producing these standard wave shapes provides a learning experience that is quite helpful to the beginner as well as the experienced additive synthesist. It should also be remembered that additive synthesis is capable of so much more than producing traditional wave shapes, so creating the standard waveforms should be used as a jumping-off point to the wonderful world of additive synthesis.

Envelope Generators

As stated in the subtractive synthesis chapter, an envelope generator affects various parameters of the circuit it is routed to. Typically, an envelope generator will contain attack, decay, sustain, and release parameters. In additive synthesis, envelope generators are usually routed to amplifiers for both individual harmonics or for the sound as a whole. Either way, an envelope generator will affect how the sound's amplitude changes over time while a key is being depressed and once it is released. Envelope generators are key when it comes to additive synthesis in order to make the resulting timbre sound interesting and not stagnant. Early additive synthesizers did not have the functionality of envelope generators, which made them sound like novelty instruments in the lens of modern additive synthesis. When programming an additive synthesizer, if one only sets the frequencies and initial amplitudes of the upper harmonics, the resulting sound will be, well, like an organ. That is because this is exactly how organs work. So in order to make additive synths sounds more interesting, one must use envelope generators. The possibilities are endless when using envelope generators. If one sets the envelope generators of all the upper harmonics to fall shortly after a key is pressed, the resulting sound will be reminiscent of a filter closing on the sound. If the envelope generators are set to rise once a key is depressed, the resulting sound will be reminiscent of a filter opening. One might ask—if envelope generators re-create filter responses, why not just add a filter to an additive synthesizer? The answer to that valid question is that creating filter responses is only one small aspect of the power of envelope generators when used by the additive synthesist.

When using envelope generators, one can set certain higher harmonics to rise in amplitude once the fundamental and lower harmonics die out in order to create interesting evolving pads. Likewise, one can set just the higher harmonics to be heard when a key is pressed and then have those die away while the fundamental and lower harmonics rise in amplitude. The types of sounds that can be created using the envelope generators of an additive synthesizer cannot be created using any other type of synthesis format. That last statement is very important to note. In theory, all sounds created by any of the synthesis formats can be re-created using additive synthesis but an additive synthesis sound can rarely be created using a different format. It is also important to note that most modern additive synthesizers will feature an additional envelope generator that can be used to affect the sound as a whole. By using this overall envelope generator, the entire sound can be swept up, swept down, held, muted, and anything in between.

Envelope generators are not just limited to amplitude control. Certain additive synthesizers will also allow envelope control over pitch for individual or groups of harmonics. Being able to designate how a harmonic changes pitch over time can greatly influence the overall timbre of the sound. For example, when setting the envelope to change the pitch of many harmonics drastically as the sound evolves, the resulting sound will start off very uniform and slowly become more and more mangled as time goes on. Another routing many additive synthesizers offer for envelope generators is the panning of individual harmonics. Newer additive synthesizers are typically stereo instruments, meaning they have a left and right output. By using envelope generators to adjust the pan function of individual harmonics, certain aspects of the sound can shift from one speaker to another, creating extremely immersive synth sounds. A world of opportunity opens when one incorporates envelope generators into an additive synthesis patch.

LFOs

LFOs, or low frequency oscillators, are a must when it comes to additive synthesis. Many modern additive synthesizers will feature independent LFOs for each and every harmonic. LFOs are typically used as another source of amplitude modulation similar to envelope generators. However, LFOs can be used as a means of frequency modulation for individual harmonics as well. Using LFOs to modulate the pitch of various harmonics will yield small, tonal variations at more modest levels while creating intense sonic warping at more extreme levels. As stated in the subtractive synthesis chapter, LFOs are oscillators that produce frequencies at lower than audible ranges. Because of this, LFOs are capable of producing the range of standard waveforms, which can then be used as a means of control. Due to the cyclical nature of wave shapes, LFOs are able to produce repeating modulation, which can impart new and creative effects when programming an additive synthesizer. Many modern additive synthesizers will also feature one or more overall LFOs that will be able to affect the sound as a whole. These overall LFOs can be used to create vibrato and tremolo effects as well as a wealth of interesting and unique effects that are only limited by the user's imagination.

Borrowed Synthesis

Although many additive synthesis purists might scoff at this next statement, many additive synthesizers feature parameters borrowed from other synthesis formats in order to make the instrument more beneficial to the user. The inclusion of these various parameters not only makes the instrument more user-friendly, but can also be used as a marketing tool to not only differentiate the instrument from other additive synthesizers, but make it more appealing to the user that is thinking about getting into additive synthesis. Although often overlooked, having an overall envelope generator or LFO on an additive synthesizer is technically a means of incorporating borrowed synthesis because the resulting sounds could technically be achieved by only using additive synthesis rather than incorporating an overall envelope or LFO.

Filters

Many additive synthesizers, including many of the Kawai instruments, contain filters. Filters, especially resonant, low pass filters, have become synonymous with synthesis. Being able to perform a filter sweep in real time is practically a must on most synthesizers. Therefore, it was only a matter of time before additive synthesizer designers began including filters into their instruments. Any type of filter response can be achieved on an additive synthesizer by manipulating the amplitudes of the individual harmonics. For example, a low pass filter sweep can be achieved by setting the higher harmonics to fade out prior to the lower harmonics. Likewise, a high pass filter sweep can be achieved by setting the lower harmonics to fade out prior to the higher harmonics. Even more complex filter responses such as band pass sweeps and notch sweeps can be achieved. By including a filter into the instrument, however, these sounds can be achieved much faster without having to go back into submenus and designate each harmonic to change amplitude. Having a filter on an additive synthesizer comes in handy when playing live or improvising in the studio. If staying true to additive synthesis is what your heart desires, the filter can typically be switched off or just left unused.

Formant Filters

Formant filters are a type of filter that can be used to impart human speech-type effects onto a sound such as vowel sounds. Formant filters are typically found on formant synthesizers (a type of physical modeling synthesis). Formant filters are oftentimes found on modern additive synthesizers, although they are not additive in nature. Incorporating a formant filter into an additive synthesizer effectively opens up more possibilities for sound creation.

Sample Playback

Another feature many additive synthesizers will include is the ability to utilize sample playback in conjunction with the main additive synthesis engine. This functionality is more common on the workstation type of synthesizers that have many different synthesis formats, but many standalone additive synthesizers will feature some sort of sample playback. Having the ability to start with a sampled sound and then add additional harmonic content onto it is greatly beneficial to the novice or even expert synthesist. Including sample playback features on an additive synthesizer is just another way companies are making their synthesizers remain relevant in an age where any type of synth sound can be recalled with the click of a mouse.

Waveform Generators

Although all of the basic waveforms can be re-created on an additive synthesizer, many additive synthesizers will allow users the option to start with premade wave shapes, which can then be altered and expanded upon. This type of functionality helps synthesists arrive at the sound they are trying to create faster and can often be extremely beneficial to the beginner.

Noise Sources

As mentioned above, most, if not all, sound contains some form of noise that plays a role in the timbre of the overall sound. This noise can be the sound of a bow scratching on a string, a woodwind or horn player's breath, or just atmospheric noise that influences the way in which the harmonics are heard. Using the Fourier series, one, in theory, could deconstruct noise down to individual sine waves, but it would take up far too many of an additive synthesizer's sine wave generators to re-create. For this reason, many additive synthesizers will feature an onboard noise generator that allows for noise shaping through filtering. By adding in noise at small amounts, the user can more faithfully re-create sound. When used to more extreme degrees, noise will impart sonic attributes to a sound much like when it is used in subtractive synthesis.

The filter used to shape noise will typically have some sort of contour control, such as an envelope generator. By using an envelope generator for the noise-shaping filter, the user is free to designate how the noise is shaped over time. An additional envelope generator for the noise's amplitude is also typically found on an additive synthesizer in order to control how the noise rises and falls in amplitude over time. By incorporating noise into an additive synthesizer patch, sounds can be convincingly re-created.

Performance Control

Now that we have covered how sounds are created using additive synthesis, let's examine how sounds are controlled. Like with most forms of synthesis today, additive synthesizers utilize keyboards in order to trigger sound. Because most additive synthesis exists in the digital world, additive synthesizer keyboards trigger sound digitally rather than with control voltage like on analog subtractive synthesizers.

Pitch and Mod Wheels

Most additive synthesizers will feature separate pitch and modulation wheels. Similar to subtractive synthesizer keyboards, the pitch wheel will shift the overall pitch up or down while the modulation wheel will control the depth of the overall LFO. Since additive synthesis is typically digital, the modulation wheel (and sometimes pitch wheel) can be routed to a variety of different parameters. For example, one could control an onboard filter cutoff or the envelope amount of a bank of harmonics.

Data Sliders

Due to the sheer amount of controls that can be tweaked on an additive synthesizer, having an individual potentiometer or slider for each and every parameter would result in a huge,

Figure 4.11 Data slider.

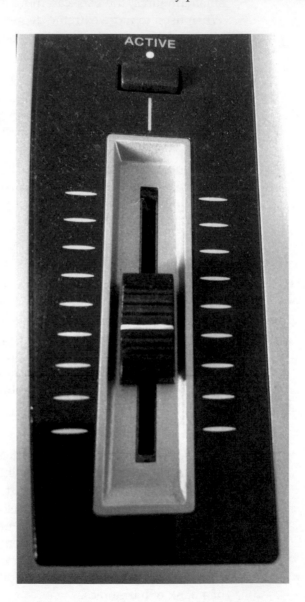

heavy, and cumbersome instrument. For this reason, most additive synthesizers will require the user to scroll through various menus and submenus on an LCD screen in order to tweak parameters. Once a parameter is selected, it will typically be altered via a data slider or bank of data sliders on the instrument's surface. A data slider typically looks and functions like a fader on a mixing console. Utilizing menus and data sliders not only allows the synthesizer to be contained in a manageably sized instrument, but allows the user to have a number of parameters routed to the data sliders at any given time in order to perform functions in real time while performing.

Sequencers/Arpeggiators

Similar to subtractive synthesizers, most additive synthesizers will feature onboard sequencers and arpeggiators. An additive synthesizer's sequencer and arpeggiators will function much the same as their subtractive synthesis counterparts, only they will almost always be controlled via digital means such as software or MIDI data.

DSP Effects

Although DSP, or digital signal processing, has nothing to do with synthesis itself, most modern synthesizers, especially digital and software synths, will feature DSP effects in order to further modify a sound. DSP effects are wide ranging and can include delay, reverb, phase shifting, chorus, and flange effects. Due to the popularity of including DSP effects on a synthesizer, most new additive synthesizers will feature at least a few DSP effects. Adding a delay or a rich reverb onto a synth patch will impart a new dimension of animated sound that, in most cases, is extremely desirable.

Resurgence of Additive Synthesis

Additive synthesis has always been more of an esoteric synthesis format. Although it is one of the earliest theorized synthesis formats as well as an extremely powerful one, additive synthesis has never had the wide acclaim that has been granted to subtractive, sampling, or even FM synthesis. This is perhaps due to the amount of time it takes not only to build a patch but to alter it on the fly. Having one of the steepest learning curves also hasn't helped additive synthesis. Sampling synthesizers single-handedly made sound re-creation via additive synthesis obsolete since any sound could be recorded and then played on a sampling synth rather than painstakingly re-created on an additive device. Once accurate sound re-creation was cast aside, many users felt that using subtractive or FM synthesis to create new sounds was sufficient enough, leaving additive synthesis behind. Although additive synthesis has never died away in the eyes of additive enthusiasts, sound designers, and academic institutions, it seemed all but dead before it came into its own for the average synthesist.

With the advent of software synthesizers and the ever-increasing amount of CPU in today's computers, however, additive synthesis is finally having its long-awaited day. Additive synthesis has finally found a platform in software synthesizers that allows it to be realized to its utmost potential.

Native-Instrument's Razor software synth is an additive synthesizer that is capable of producing up to 320 harmonics. Take into consideration that the Kawai K5, arguably the most famous hardware additive synthesizer of all time, could

Figure 4.12 Native-Instrument's Razor synthesizer.

only create 64 harmonics or 128 if one was ok with losing four notes of polyphony. The fact that Razor can produce 320 harmonics while maintaining full polyphony is nothing short of amazing.

Many software synth manufacturers are including additive synthesizers or at least elements of additive synthesis into their new products. It is an exciting time in the additive synthesis world. As CPU power increases, the limits of additive synthesis will continue to expand resulting in capabilities much beyond our current understandings.

It is appropriate to say that software is the perfect additive synthesis format. Not only will a software additive synth be much cheaper in cost than its hardware counterpart, but it will also be much more powerful. In fact, software additive synthesizers are free to be expanded and built on with subsequent updates keeping them relevant and state of the art so long as the software company stays in business. This is an extremely important note because it means that once a user buys a software additive synth, it can potentially be kept up to date with every advancement in technology for the small cost of updating the software. Updating a synthesizer as technology improves can really only be possible through the means of software synths.

Like Native-Instrument's Razor synthesizer, AIR has its own software additive synthesizer called Loom. Loom can

Figure 4.13 AIR Loom synthesizer.

be considered even more powerful than Razor because it offers the user 512 individual harmonics. The trend in software additive synthesis to feature as many harmonics as possible with average CPU power is both exciting and amazing. Comparing the power of Razor or Loom to the likes of the Kawai K5 show that software is the next logical evolution for additive synthesis thanks to the sheer power it affords.

Utilizing software for additive synthesis not only allows for huge numbers of harmonic generators, but allows for huge numbers of individual amplitude control as well. Returning back to the Kawai K5, only a select number of envelope generators were available. Users would have to assign individual envelope generators to a group of harmonics. Although certain hardware additive synthesizers featured independent envelope generators for each harmonic, it was not the norm. Software additive synthesizers, on the other hand, almost exclusively offer individual envelope generators for each harmonic. Many software additive synths go as far as to add more parameters than the standard attack, decay, sustain, and release into their envelope generators. Some software synthesizers even include multiple envelope generators per harmonic. For example, Camel Audio's Alchemy Synthesizer features 600 individual harmonics and includes three envelope generators for each harmonic. When using Alchemy, the user has a designated envelope for each harmonic's amplitude, pitch, and panning. Alchemy is an additive synth that can only exist in the software world due to its sheer power. Camel Audio's Alchemy is a great example of the new era of additive synthesis. Not only does Alchemy sport one of the most powerful additive synthesis engines available, but it also has a designated subtractive synthesizer and fully featured sample-based and granular synthesizers as well. Alchemy, like many other modern software synthesizers, is taking advantage of CPU power in order to give the user the most sonic possibilities available for sound creation.

Recipes

No two additive synthesizers are exactly alike. The number of available tone generators, the way in which partials or harmonics are laid out, and the ways in which they can be manipulated will be unique to each instrument. Because of this, simply drawing out recipes for various patches will not work because, chances are, an additive synthesizer that you are using to re-create these patches will not have the same features as the additive synth we are using to create these patches. Instead, we have created ten patches that we feel show the capabilities of additive synthesis with an emphasis on patches that stray away from the stereotypical additive pad type sounds. We will then demonstrate how we created each of these patches on our given soft synth with plenty of descriptions and figures.

Each of these ten patches were created using the soft synth Loom by AIR. Loom was chosen for not only its popularity as an additive synthesizer, but for its capabilities as well. In the following pages, we will walk you through the creation of each of these patches in an attempt to help demystify some of the functions of additive synthesis. Before we delve into the patches themselves, let's go over the various parameters of Loom and what each function does.

Loom

Loom is a software additive synthesizer that is laid out in a modular manner. In the edit page of the synth, there are ten spaces where users can select which function module they would like to put. Each subsequent module will affect the sound that is coming from the previous module. In addition to the function module, there are some overall attributes that can be affected as well as a fairly cool *Morph* page which allows for point-and-click sound morphing. Let's start with exploring the edit page and the various function modules.

Once Loom is launched, we click the edit tab in the upper left corner and then we can load the main sound creation page. In

Figure 4.14 Loom's default blank screen.

the upper part of the screen, we have the ability to select the number of partials present per voice. Loom has the ability to produce up to 512 partials per voice. Moving down the page, we see ten blank spaces, which we can load sound and function modules into. As stated before, each module is fed into the next, so it's important to think about where in the chain each sound module should be placed. By clicking the bar at the top of each module space, we can select which function module we would like to insert. Let's look at each of the modules to know what they do.

Gain

The *Gain* module acts like an amplifier in that it controls the volume of all the partials together. A *Gain* module is needed at some point in the chain in order to designate partial volume.

Odd/Even

The *Odd/Even* module is one of the most crucial modules in Loom. The module controls the level of the fundamental frequency and all partials. The *Odd/Even* module also allows you to change the level relationship between the odd and even harmonics in order to change the timbre of the overall sound. Finally, a damping control is available that dampens higher harmonics, giving the user just one more avenue of sonic manipulation.

Figure 4.15 Loom *Gain* module.

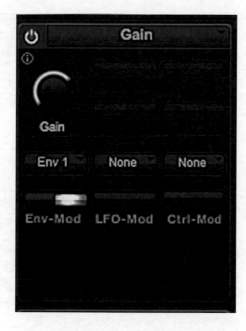

Figure 4.16 Loom *Odd/Even* module.

Figure 4.17 Loom *Sync* module.

Figure 4.17 Loom *Sync* module.

Sync

The *Sync* module changes the partial relationships in a way that attempts to mimic an oscillator sync type sound on a subtractive synthesizer. The user has the ability to change the frequency of the imaginary second oscillator as well as the depth of the effect. Finally, the ability to smooth the effect as a whole is present. It's important to note that this module is not truly adding a second oscillator and then syncing it; the module is simply changing the tuning and levels of individual harmonics in order to mimic a synced sound.

Second Tone

The *Second Tone* module is another module that changes the relationship of harmonics. The module is described as one that attempts to mimic the sonic changes inherent in a subtractive synthesizer when another oscillator is added into the mix. However, in practice, the module sounds more like it is putting more focus onto another fundamental frequency,

Figure 4.18 Loom *Second Tone* module.

which allows the sound to change drastically. By detuning the *Second Tone* module, extremely interesting harmonics begin to become present.

Octaver

The *Octaver* module is a fairly simple module which creates copies of the original sound spectrum and shifts their octave. The user has the ability to control the depth of the effect as well as the number of copies and octaves that are made. Finally, a damping control is available with which to lessen the effect.

Organ I and II

The organ modules are simple modules that just remove certain harmonics in an attempt to create a sound which is akin to an organ. The difference between *Organ I* and *Organ II* is simply the harmonics they remove.

Figure 4.19 Loom *Octaver* module.

Figure 4.20 Loom *Organ I* and *II* modules.

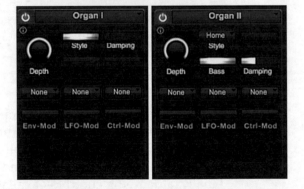

Discrete

The *Discrete* module is similar to the organ modules in that it removes certain harmonics in an attempt to mimic certain sounds. In the case of the *Discrete* module, harmonics are removed in order to mimic the spectrum of a bell or drum.

Figure 4.21 Loom *Discrete*
module.

Filtering Modules

Loom features a series of "filtering" modules that attempt to mimic the sound of various filter shapes. Although many of you may be screaming that filtering is not an additive synthesis trait, Loom's filtering modules are not exactly filters in and of themselves. The filtering modules in Loom mimic traditional filter shapes through additive synthesis by removing harmonics, changing the relationships of harmonics, and adding harmonics in the case of mimicking resonance.

Figure 4.22 Loom filtering
modules.

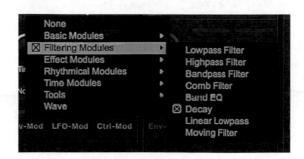

Figure 4.23 Loom decay module.

Decay

Located under the filtering tab, *Decay* is a pretty handy module that should not be overlooked. The decay module allows users to determine how individual partials will fade out. There's a control that determines the amount of time it takes for the entire sound spectrum to fade out as well as a control to designate whether higher or lower partials will fade out first.

Effect Modules

Loom features a series of effect modules that each have specific tasks. In the lineup, is a *Detune* module, which changes the levels of partials in an attempt to mimic the way a subtractive synthesizer sounds when a number of oscillators are detuned from each other. Next is a *Modulator* that modulates the levels of various partials in a random manner. Finally there are *Phaser*, *Stereo*, *Pan*, and *Blur* modules. The *Phaser*

Figure 4.24 Loom effect modules.

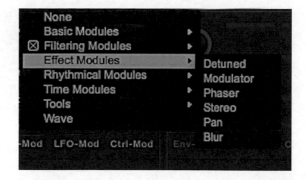

adds a phaser-type sound, whereas the *Stereo* and *Pan* modules allow users to designate how the sound will fit in the stereo field. Finally, the *Blur* module affects individual harmonic volumes randomly at various degrees in an attempt to blur the sound.

Phrase and Random Drops

The *Phrase* module increases the level of individual harmonics using predefined patterns inside the software. Users can scroll the many rhythmic patterns inside the software. The *Random Drops* module effectively does the same thing as the *Phrase* module, but in a random manner rather than set to a rhythmic pattern.

Figure 4.25 Loom *Phrase* and *Random Drops* modules.

Figure 4.26 Loom time modules.

Time Modules

The two time modules—*Tail* and *Repeater*—are designed to mimic specific effects using additive synthesis. *Tail* adjusts partial levels to create reverb like effects while *Repeater* does the same thing to create a delay type effect.

Tool Modules

There are a variety of modules in the tool category. Each of them has a small, but specific purpose. The *Adder* module adds a constant level to specific harmonic even after it might have been filtered out while the *Constant* module does the same thing to a set of harmonics. The *Enhancer* module adjusts the overall sound much like a sonic enhancer would. The *Clip* module adds a pseudo-distortion type effect. The *Threshold* module can act like a compressor or limiter and, finally, the *Noise* module adjusts harmonic levels to create what sounds like noise.

Figure 4.27 Loom tool modules.

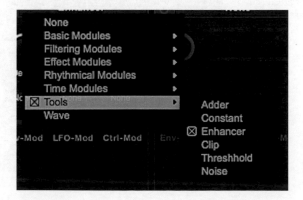

Figure 4.28 Loom *Wave* module.

Wave

The *Wave* module is actually a really cool module. It's not specifically an additive synthesis module, but it's a great addition. Using the *Wave* module, users can import audio clips of their choosing, which can then affect the levels of all partials like a vocoder. There are a bunch of controls to customize the sound, but some really interesting effects can be generated using this module.

Globals

Now that we have finished discussing individual modules, let's take a look at some of the global parameters which can be assigned to affect each module. At the bottom of Loom's edit screen, we are greeted with a variety of parameters, such as

Figure 4.29 Loom globals.

157

envelope generators, LFOs, and pitch control. The envelopes and LFOs can all be routed to individual parameters of the aforementioned modules in order to add modulation to each module while the global pitch parameter allows us to have the envelopes and LFOs affect the pitch of the synthesizer as a whole. Loom features three AHDSR (attack, hold, decay, sustain, release) envelope generators as well as a fourth slope envelope generator. The three LFOs have a wide variety of waveforms including sine, square, saw up, saw down, and sample and hold. Two additional LFO responses are available, pump and drift, which create weird modulation shapes that are quite fun to play with.

FX and Master

Directly above and to the right of the global parameters are two small sections entitled *FX* and *Master*. As can be guessed, the *FX* section allows the user to add an overall effect to the sound, such as distortion or reverb, while the *Master* section acts as an overall level and stereo width control.

Figure 4.30 Loom *FX* and *Master* sections.

Figure 4.31 Loom *Spectral Modulation/Distortion* sections.

Spectral Modulation/Distortion

In the top left corner of the edit screen are two sections entitled *Spectral Distortion* and *Spectral Modulation*. The *Spectral Distortion* section adjusts algorithms inside the software to affect the frequencies of the partials. The *Spectral Modulation* section allows for modulation of the partial frequencies via LFO or envelopes. Both these controls are unlike the individual modules previously discussed. These two sections both affect harmonic frequency rather than harmonic level.

Morph Screen

The morph pad in the morph screen acts sort of like a vector plane on a vector synthesizer. Users can place four variations of the sound at each quadrant and draw paths for the sound to move through at set speeds and directions. The morph screen is quite cool and allows for some really extraordinary movement and modulation options. The morph control is pretty deep and is out of the scope of additive synthesis, so

Figure 4.32 Loom morph screen.

we won't go into it in depth. Just imagine it as a glorified vector plane and control.

We have now covered the various parameters in AIR's Loom additive synthesizer. Now that you have a brief understanding of how the synthesizer works as well as what it can do, let's examine each of the additive patches we have created and hopefully they will serve as jumping-off points for you to explore the wonderful world of additive synthesis.

Recipe 1: Ambient Bass

The first patch we created is a take on the synth bass sounds heard in electronic ambient music. The patch features 192 partials and begins with an *Odd/Even* module with the fundamental fully cranked and a slight amount of damping on the harmonics. The Odd/Even knob is turned so more even harmonics are heard than odd harmonics. Next in the chain is a *Discrete* module, which has its depth turned fairly high. The brightness, fundamental, random, and spectrum controls are also turned fairly high allowing for some weird sonic attributes. After the *Discrete* module, a *Second Tone* module is placed with full detune to add more complexity to the sound. After that, a small amount of *Blur* and *Decay* are added to smooth out the sound as well as let the low harmonics ring out longer rather than the high. Finally, a *Tail* and *Gain* module are placed to add some reverb type sound as well as

Figure 4.33 Screenshot of Ambient Bass patch.

control the level of the sound as a whole. The overall pitch of the patch is being slightly affected by LFO one, which is set to a sine wave and beat-synced. A slightly driven distortion is placed on the sound as well as a 50% reverb setting. This patch at first sounds like a weird natural distorted bass and then gets more and more mangled as the sound fades out.

Recipe 2: Cool Repeater

The second recipe in our list plays on the use of additive synthesis in creating a delay-type effect. The sound features 256 partials and starts with an *Odd/Even* module with only a slight amount of the fundamental frequency added in. The harmonics are fairly damped and mixed in order for both

Figure 4.34 Screenshot of Cool Repeater patch.

odd and even harmonics to be heard at the same level. Next, a *Detune* module is placed in the chain with two stages of detuning and only a slight detune. After the *Detune* module, a *Low Pass Filter* module is placed fully "open" with a bit of resonance and a large amount of envelope modulation. Next, an *Enhancer* module is placed with just a slight amount of the effect being heard. Finally, a *Gain* and *Tail* modules are placed at the end of the chain. An important note is that the *Tail* module is placed after the *Gain* module in order to make the reverb effect still ring out once the sound has ceased. Finally, distortion, reverb, and delay are added in the *Master* section of the synth.

Recipe 3: Distorted Bells

The "Distorted Bells" patch we created was designed so that when a note is first played, the bell tone is fairly clear. Once the notes start ringing out and overlapping, however, the distortion begins to emerge and change the sound drastically. The patch contains 256 partials and is made up of an *Odd/Even* module with a bit more emphasis on the odd harmonics. A bit of damping is added while the fundamental is turned about half way up. Next, a *Discrete* module is added in order to achieve the bell sound that we were going for. A *Band EQ* filter module is placed with a small amount of gain and a bit of key tracking just to smooth out the sound and let it

Figure 4.35 Screenshot of Distorted Bells patch.

evolve. Next, *Blur*, *Decay*, and *Tail* modules are added not only to smooth the sound, but to add a fairly long ring out that becomes darker as it progresses. Finally, a *Gain* module is added for level control. In the *Master* section, distortion and delay are added as well as the overall pitch being modulated by envelope four.

Recipe 4: Distorted Movement

The "Distorted Movement" patch we created was designed to fully utilize the type of digital grit that can be achieved using an additive synthesizer. The sound features 192 partials and starts with an *Odd/Even* module with a small amount of damping and hardly any fundamental present. Again, more emphasis is placed on the odd harmonics. Next, a *Decay* module is added in order to let the sound ring out. A *Gain* module is placed directly after these to allow for level control over just this aspect of the sound. After the *Gain* module is a *Sync* module, which adds a lot of the digital grit we were looking for. A *Stereo* module is then added with max depth to influence the sound's presence in the stereo field. Directly after the *Stereo* module is a *Low Pass Filter* module with heavy influence from its respective envelope and LFO. The frequency control is slightly rolled back on the *Low Pass Filter* module as well. Finally, the sound chain ends with another *Gain* module in order to control the level of the sound as a whole. Distortion,

Figure 4.36 Screenshot of Distorted Movement patch.

delay, and reverb are all added in the *Master* section as well as a slight amount of envelope four modulating the overall pitch.

Recipe 5: Electric Guitar Lead

This next patch was not originally created with an electric guitar sound in mind, but as it progressed, it was evident that an electric guitar is what sprang to mind. The patch features 256 partials and begins with an *Odd/Even* module with a fair amount of the fundamental added in as well as a good amount of harmonic damping. The Odd/Even knob is left in the middle with equal levels for both odd and even harmonics. Directly after the *Odd/Even* module are *Low Pass Filter* and *High Pass Filter* modules. The *Low Pass Filter* module's frequency is left fairly high with a large amount of key tracking, while the *High Pass Filter* module's frequency is rolled further back and an extreme amount of resonance added. After the two filter modules, *Decay* and *Tail* modules are added to affect how the sound rings out. Next, an *Enhancer* module is added with a large amount of depth followed by a *Gain* module to control the overall level of the sound. Finally, distortion and reverb are added in the *Master* section. The patch itself is fairly straightforward and sounds menacing.

Figure 4.37 Screenshot of Electric Guitar Lead patch.

Recipe 6: Low String Repeat

The "Low String Repeat" patch we created is pretty interesting. The initial sound is akin to natural strings played in a pizzicato manner, but then LFO-caused repeats are heard from select harmonics making for a strange delay-type of sound that almost appears to be a completely different instrument. This patch features 192 partials and begins with an *Odd/Even* module with full emphasis on the odd harmonics. The fundamental is almost entirely left out and a bit of harmonic damping is added. Next a *Discrete* module with a heavy depth and a good amount of brightness is added, which plays heavily in the repeating aspect of the sound. Next in the chain is a *Second Tone* module with full level and a bit of damping, which helps not only the pizzicato string aspect of the sound, but the repeating aspect as well. Next there are *Blur*, *Decay*, and *Tail* modules, which will heavily affect the repeating aspect of the sound due to the sharp decline of the pizzicato sound. Finally, a *Gain* module is added for overall harmonic level control. Distortion and reverb are added in the *Master* section as well as a slight pitch modulation from envelope four. LFO three is set to a square wave with retrigger initiated: it is controlling the *Spectral Modulation* section of the synthesizer that is causing the repeated sound.

Figure 4.38 Screenshot of Low String Repeat patch.

Figure 4.39 Screenshot of Natural Low Strings patch.

Recipe 7: Natural Low Strings

The "Natural Low Strings" patch we created was designed to sound like a heavily affected recording of some type of stringed instrument. The patch features 128 partials and begins with an *Odd/Even* module. The fundamental is fully cranked with about half damping and equal representation of odd and even harmonics. Next, a *Band EQ* module is added with a midrange frequency. A small amount of key tracking is enabled as well as slight envelope modulation. Next, a *Low Pass Filter* module is added with the frequency knob turned up about 75%. *Pan*, *Blur*, and *Decay* modules are then added to affect the sound's tail as well as presence in the stereo field. Finally, *Moving Filter* and *Gain* modules are added. The *Moving Filter* module is set to "stop" but can be initiated as desired. Delay and reverb are then added at the *Master* stage.

Recipe 8: Metallic Sequencer

The "Metallic Sequencer" patch is designed as a digital-heavy patch that lends itself well to a step sequencer. The patch features 256 partials and begins with an *Odd/Even* module with little damping and fundamental present. Heavy emphasis is placed on the odd harmonics. Next, a *Gain* module is added in order to have control over just the sound coming from the *Odd/Even* module. Next, a *Repeater* module is added with a

Figure 4.40 Screenshot of Metallic Sequencer patch.

low level. Directly after the *Repeater* is a *Low Pass Filter* module with its frequency turned all the way up and a bit of resonance as well as heavy emphasis width. Next, *Stereo* and *Blur* modules are placed in order to affect the sound as a whole as well as its place in the stereo field. Another *Low Pass Filter* module is added to darken the sound a bit as it is quite harsh at this stage. The *Low Pass Filter* module is placed after the *Stereo* and *Blur* modules to ensure it is darkening the sound emitting from these modules. Finally, a *Gain* module is added to control the level of the entire sound. Reverb is then added in the *Master* section.

Recipe 9: Moving Bell

The "Moving Bell" patch is a cool, sonically evolving bell sound that sounds quite different from any bell sound we had heard. The patch contains 128 partials and begins with an *Organ I* module with a heavy amount of depth and damping. Directly after the *Organ I* module, a *Discrete* module is placed, which aids in getting the bell sound required for this patch. Next, two *Low Pass Filter* modules are added with different cutoffs in order to give the sound two stages of filtering. Next, a *Gain* module is added in order to have control over the sound at this point. An *Adder* module is then added with a fairly high frequency in order to add some high harmonic artifacts back into the sound. Next, *Stereo* and *Tail* modules

Figure 4.41 Screenshot of Moving Bell patch.

are added in order to allow the sound to fit nicely in the stereo field and ring out. A *Phrase* module is then added with the "oct-1" pattern. Finally, a *Gain* module is added for overall control. Delay and reverb are both added in the *Master* section as well as slight pitch modulation from envelope four. Envelope one is set in a trapezoid type shape and is slightly modulating the *Spectral Modulation* section of the synth.

Recipe 10: Synth Bass

The Synth Bass patch we created was designed to mimic old, acid-style synth basses with a bit of additive flair. The sound features 192 partials and begins with an *Odd/Even* module with heavy emphasis on odd harmonics as well as a strong fundamental. The sound moves on to feature both *Discrete*

Figure 4.42 Screenshot of Synth Bass patch.

and *Second Tone* modules, which help achieve some of the piercing "squeakiness" present in acid bass sounds. Next, *Blur*, *Decay*, and *Tail* modules are added to affect how the sound rings out. A *Gain* module is then added for level control. Distortion and reverb are then added in the master stage.

Historical Perspective on Additive Synthesis

One of the earliest instruments to incorporate additive synthesis into its sound creation was the Teleharmonium.[5] The Teleharmonium was invented by Thaddeus Cahill in 1897. Cahill, who studied physics of music at Oberlin Conservatory in Ohio, was consumed with the idea that music could be made via electronic means. While working on the designs for an electronic typewriter, Cahill began work on the Teleharmonium. The Teleharmonium was originally designed as a device that could play music over telephone lines. Cahill was hoping to market the Teleharmonium to hotels and restaurants in order to have one instrument feeding music to a variety of places at the same time. Cahill hoped the Teleharmonium would convince hotel and restaurant owners to stop paying pianists and small string ensembles and instead pay him a lower cost to produce music at their venue via telephone lines. During the time Cahill was working on the Teleharmonium, however, there was no way in which to amplify sound coming from a telephone's headset. Cahill surmised that if his device created enough electricity, it would create a loud enough signal that could then be fed from a telephone's earpiece into an elongated cone, much like a gramophone cone, causing the signal to be heard by an audience. Cahill's early work on sound amplification also helped pave the way for loudspeaker design.[6]

The Teleharmonium was an extremely large contraption that took up at least two rooms and, in some instances, weighed upwards of two hundred tons. The first Teleharmonium Cahill built was a scaled-down version of his original plan and incorporated 35 elongated cylinders that contained a number of tone wheels each. The tone wheels on the Tele-harmonium were a series of wheels that had several raised bumps on their surface. The tone wheel cylinder would then spin while a magnetic coil was placed close to the bumps. When a bump passed by the magnetic coil, electricity was generated. The space between the bumps generated little or no electricity. This alternating current, caused by the bumps on the tone wheel, is what produced sound. The cylinder was divided into a number of sections that had a different number of raised bumps, causing different pitches to be created. The sections of the cylinder were capable of rotating at a various speeds, resulting in a wide range of frequencies. The various tone wheels could then be combined in order to produce rich, pleasing tones.[7]

The second Teleharmonium that Cahill built was much larger and contained 145 tone wheel cylinders, resulting in much more complex tones that could be generated. Cahill built a final, third Teleharmonium and displayed it in 1910. This final Teleharmonium was again larger than its two predecessors, but the novelty of the instrument had worn off and the public had grown tired of it. The diminished interest as well as the sheer weight and size of the Teleharmonium caused its ultimate downfall.[8] However, the Teleharmonium secured its place in history not only as the first machine to use additive synthesis, but the first synthetic instrument.

Hammond Tone Wheel Organ

Following in the footsteps of Thaddeus Cahill, Laurens Hammond and John M. Hanert manufactured the Hammond tone wheel organ. Like the Teleharmonium, the Hammond organ utilized tone wheels. The tone wheels on the Hammond, however, were small discs that had designated electronic pickups in close proximity to the disc. Alternating Current (AC) motors were responsible for spinning the discs at exact speeds. The speed at which the tone wheels spun, as well as the number of physical bumps on their surface, determined the frequency that would be produced.[9]

The Hammond organ was unique in that it had a series of controls called drawbars that the user could engage in order to produce higher harmonics. If only one draw bar was engaged, the organ would produce a close proximity of a sine wave. The sine wave would produce a frequency corresponding to the note being held down on the keyboard. This first draw bar represented the first harmonic, or fundamental frequency. Engaging subsequent drawbars would introduce higher harmonics into the sound with each draw bar representing the next harmonic. The user was able to control the amplitude of the various harmonics by physically pulling the drawbar out further or pushing it back in, resulting in louder and softer harmonics respectively.[10]

Being able to incorporate individual harmonics into a sound, as well as their amplitude, is the fundamental concept of additive synthesis. However, the Hammond organ, as well as the Teleharmonium before it, was incapable of changing the amplitude of these higher harmonics overtime, which is another key element of successful additive synthesis.

Early Electronic Additive Synthesizers

Although the Hammond organ and the Teleharmonium used additive synthesis in order to generate sound, they did so via mechanical means. Both instruments incorporated tone wheels that physically spun in order to create sound. Additive synthesis produced through electronic means would be the next evolution for this synthesis format. One of the earliest all-electronic additive synthesizers was produced by E. L. Kent in 1942, an experimental machine that helped drive early research into electronic additive synthesis.[11]

Building on the research of E. L. Kent, James Beauchamp, working out of the University of Illinois, invented and built the first practical electronic additive synthesizer in 1964, known as the Harmonic Tone Generator. The Harmonic Tone Generator was a relatively small (compared to the monstrous Teleharmonium) electronic instrument that was capable of producing up to six exact harmonics with a varying frequency range up to 2,000Hz.[12] The unique thing about the Harmonic Tone Generator, besides its all-electronic construction, was the fact that the individual harmonic's amplitude and phase relationships could be controlled. The Harmonic Tone Generator used control voltage (the same technology referred to in the subtractive synthesis chapter) in order to control the various harmonic attributes. The ability to control the amplitude of individual harmonics was no small matter. By simply incorporating harmonic amplitude control in the time realm, additive synthesis could finally become recognized as the powerful synthesis technique that it is.

Prior to the Harmonic Tone Generator, instruments that incorporated additive synthesis only allowed the initial amplitude of individual harmonics to be set. This means that the tone produced, although often interesting and pleasing, was stagnant. Natural sounds, as you may know, are rarely stagnant; they have a movement to them that allows them to evolve and remain interesting. The fact that harmonic amplitudes could be controlled via control voltage in the Harmonic Tone Generator meant that, finally, interesting and moving sounds could be produced.

The Harmonic Tone Generator used LFOs and envelope generators as a source of control for the individual harmonics.[13] By setting a slow attack on individual, higher order harmonics, the sound produced would grow more complex as time went on and as richer sonorities of harmonics became audible. Using LFOs would result in ever-changing sonorities of harmonics. Although the Harmonic Tone Generator was revolutionary, it was limited to the fact that it could only produce up to six harmonics. Being that the Harmonic Tone Generator used analog circuitry, it was also susceptible to the common problems of analog in that the pitches were unstable and tended to drift, which proved more troublesome in an additive synthesizer than a subtractive one.

The next evolution of additive synthesis would come once the digital revolution took over synthesizers. Although a few analog subtractive synthesizers incorporated some additive elements into their design, a practical, full-fledged additive synthesizer was still not available to the masses.

Arguably the first commercially successful additive synthesizer was the New England Digital Synclavier II. The Synclavier II was a state-of-the-art sampler/workstation that was released in 1980. Although typically thought of solely as a sampler, the Synclavier II featured an advanced additive resynthesis engine. The Synclavier II was extremely expensive, often costing upwards of 50 thousand dollars, and therefore was not adopted by the average musician and synthesist. The first steps into producing an affordable, fully featured additive synthesizer would be in the form of the DK Synergy and Kurzweil K150 synthesizers. Both of these synthesizers offered the user extremely powerful additive synthesis engines with the Kurzweil K150 boasting an impressive 240 oscillators!

Although the K150 and Synergy synthesizers were powerful and much less expensive than the New England Digital Synclavier II, they never caught on to the degree as other synthesizers utilizing different formats that were released around the same time. This is due in part because of the amount of time it took to program a patch as well as the steeper learning curve that came along with additive synthesis. As stated above, it would not be until the Kawai K5 was released that additive synthesis would finally fall into the hands of the average synthesist.

Additive synthesis, although old in theory, is still evolving every day. The limits of additive synthesis are yet to be explored. Beginning with the Teleharmonium and Hammond organ, additive synthesis evolved to utilize electronic sound generation in the Harmonic Tone Generator. As the digital age dawned, additive synthesis was on the forefront of the technology with instruments like the Kawai K5. As digital hardware synths moved into virtual software synths, additive synthesis followed and became even more powerful. It seems that every new additive synthesizer released has more features and harmonics available to the user. Additive synthesis will continue to grow and become even more powerful. Although additive synthesis might never become a widely adopted synthesis format like subtractive or FM synthesis, it has made its mark and continues to do so amongst the most die-hard of synthesis enthusiasts. I urge you to experiment with an additive synthesizer and see the extreme potential this synthesis format is capable of.

Notes

1. Paul Wiffen, "Synth School, Part 4: Additive Synthesis," *Sound on Sound* (October 1997). Retrieved from https://www.soundon sound.com/sos/1997_articles/oct97/synthschool4.html
2. Paul Wiffen, "Synth School, Part 4: Additive Synthesis."
3. Paul Wiffen, "Synth School, Part 4: Additive Synthesis."
4. Amar Chaudhary, "Band-Limited Simulation of Analog Synthesizer Modules by Additive Synthesis." Paper presented at the annual *Audio Engineering Society Convention*. September 26–29, 1998, San Francisco, California.
5. Robert A. Moog, "Electronic Music," *Journal of the Audio Engineering Society* vol. 25, no. 10/11 (November 1, 1997), pp. 855–861.
6. Jay Williston, "Thaddeus Cahill's Teleharmonium." Retrieved from http://www.synthmuseum.com/magazine/0102jw.html
7. Jay Williston, "Thaddeus Cahill's Teleharmonium."
8. Jay Williston, "Thaddeus Cahill's Teleharmonium."
9. J. W. Beauchamp, "Additive Synthesis of Harmonic Musical Tones," *Journal of the Audio Engineering Society* vol. 14, no. 4 (October 1966), pp. 332–342.
10. Robert A. Moog, "Electronic Music."
11. J. W. Beauchamp, "Additive Synthesis of Harmonic Musical Tones."
12. J. W. Beauchamp, "Additive Synthesis of Harmonic Musical Tones."
13. J. W. Beauchamp, "Additive Synthesis of Harmonic Musical Tones."

WAVETABLE SYNTHESIS 5

Wavetable synthesis is perhaps the most unique type of synthesis and is like a chameleon that can imitate other types of synthesis. It combines elements of sampling, vector synthesis, and subtractive synthesis, all in a system that can produce a range of sounds from subtle tones to otherworldly soundscapes. At the heart is a wavetable that has many individual waveforms, and the familiar sound of wavetable synthesis comes from playing them in order, front to back, back to front, or in any manner the available modulators can manage at a variety of speeds.

The concept of wavetable synthesis was created by Wolfgang Palm of PPG (Palm Products GmbH) in the years just before MIDI was invented: his original instruments, called PPG Wave, are still sought after today. The PPG Wave series (2.0–2.3) spanned the 1980s and, as has happened again and again, they couldn't survive in a market that brought cheaper and cheaper instruments. The good news is that other companies such as Waldorf picked up where the PPG Wave left off and there are some excellent wavetable synths currently available.

Figure 5.1 PPG Wave 2.3.

Basic Principles

The wavetable is a collection of single cycle sounds that are accessible for manipulation. A common example is a set of very similar waveforms that, if played sequentially, would sound like a sawtooth waveform that is being sent through a filter that is opening up. The wavetable process allows the sequence of files to be played back slowly, quickly, or back and forth with an LFO. Sweeping through waveforms can sound very organic if the waves are very similar, and when a synth uses interpolation techniques to mix from one to the next, it can potentially sound like an independent waveform. If the individual waveforms are varied and not consistent from one to the next, then they can still be swept through, but it will sound quite surprising and you will hear the steps between them.

How many waveforms are available in wavetable synthesis? It varies from instrument to instrument, but early PPG instruments had hundreds of individual waveforms, organized into

Figure 5.2 Wavetable instrument codex.

30 wavetables with 64 individual waveforms in each. One of the most recent wavetable synths called Nave from Waldorf has 84 wavetables, some with nearly 400 individual waveforms. There isn't a standard specification for which types of tables and how many waveforms, and so each instrument has their own individual options.

Once a wavetable is loaded, there are many adjustable parameters for how it will sound when performed. The terms used in this section are taken from the standard naming convention of Waldorf, which continues to be the primary contributor to wavetable synthesis.

Travel

The simplest movement through the wavetable is forward or backward and this directionality and its accompanying speed are determined by the travel parameter. When you hold a note with a travel setting that is set to positive at 5–10%, then the wavetable slowly cycles in a forward direction through all of the sounds. If the travel setting is set to negative at 90–100%, then the wavetable very quickly cycles through all of the sounds in reverse. At higher values, the travel settings are capable of FM-like sounds as the cycling begins to oscillate at audible frequencies.

Figure 5.3 Wavetable movement.

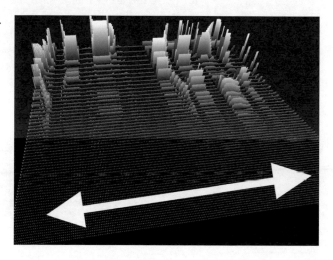

The starting point is the first thing to set because that determines where the travel begins. Some instruments use looping instead of travel, and they have different options for moving forward/backward through the wavetable. Locating the best start position is critical to achieving different sounds, and deciding which direction to move through the wavetable can change the entire sound of the patch.

Wave Modulation

The ability to modulate the playback of waveforms is at the heart of wavetable synthesis, and so most instruments that specialize in wavetable synthesis are going to have a lot of options. Here are a few of the most interesting options.

1. Key tracking—This modulation source maps the individual sounds across your MIDI keyboard, which creates an instrument that morphs depending on the range it is played in. A patch might have thick, round, lower notes, but it will morph into an edgy, screaming lead several octaves higher. The strength of this technique is that it makes performing with a wavetable very intuitive.
2. LFO—A low frequency oscillator is a classic modulation tool that is able to add a slightly different variation to what the travel option is able to provide. The results end up in a similar vein, with a cyclic pattern moving through the wavetable. If the LFO has a sample and hold option, then it is even more flexible and can create random timing cycles.
3. Envelope—An envelope is able to create motion through the wavetable in a very calculated manner, which involves various speeds through the envelope time periods. If this is applied to wavetables that involve traditional synth

Figure 5.4 Modulation matrix.

waveforms then it can readily mimic the evolution of sound over time.

4. Velocity—Using velocity to move through the wavetable is another elegant way to make performing with a wavetable more intuitive. In the same way that creating additional velocity layers in a sampled instrument adds to its overall realism, a wavetable that is mapped to velocity is probably as close to a natural instrument as any synth can get because it often involves more than 60 levels of changing timbres.

5. Modulation Pads—An X-Y pad is effective at controlling the wavetable, but it means you have to limit your playing to one hand while the other performs the modulation. The benefit to a pad is that it can control multiple parameters in addition to the wavetable location.

6. Modulation Wheel—The benefit to using a mod wheel is that many performers are already used to playing with one hand while using a wheel, but the biggest limitation is that it only modulates one parameter while limiting you to one hand.

As you can see, there are many ways to control the wavetable playback using modulation. None of them are the "right" way, and each one has a time and place to be used. Some wavetable synthesizers also have a spectrum control, which has the ability to be controlled by all of the same modulators, and this controls the frequency content of the wavetable just as an equalizer would.

Creating Wavetables

Wavetables can be created from any audio file and, with a little careful planning, you can harness the power and tools of wavetable synthesis. It hasn't always been easy to create custom wavetables, and the earliest instruments required additional modules. The PPG Waveterm is a rack-mounted unit that adds sampling to the PPG Wave, and it is very different from sampling as we know it today.

There were two available versions of the Waveterm called Waveterm A and Waveterm B. The Waveterm B updated the Waveterm A sampling rate from 8-bit sampling to 12-bit sampling, which is archaic by modern standards, but since wavetable synths most often mangle the original sounds, this limitation makes less of a difference. The Waveterm also has a limit of three seconds for sampling, which seems like a major shortcoming, but it is amazing what you can do with such a short sample. The best part? The Waveterm editor is essentially a text editor and relies on a 5.25-inch floppy disk system. Things have certainly come a long way.

The most recent process for creating wavetables is to import an audio file and let the instrument do the rest. Knowing how to manipulate audio files in a DAW is a plus, and as you'll see in the following examples, there are many things that have to be done outside of the wavetable synth itself.

Preparation Examples

1. The easiest method is to record your source into a single file, export it, and then import it into your instrument. The Waldorf Nave instrument lets you import from the clipboard or copy files from your computer via iTunes. The benefit to this approach is that you can focus on making a simple recording without the editing or complex signal processing, and then it is a simple transfer to your instrument. Homework: download an audio file and import it into your wavetable synthesizer. A particularly fun file that is available online is the Mac startup sound, which has a rich timbre and is fun to twist around using wavetable synthesis modulators.

Figure 5.5 Editing waveforms in Logic Pro X.

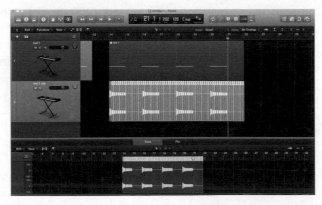

2. A more complex example involves basic editing and signal processing, which is easiest when accomplished in a DAW. Editing involves trimming the beginning and ending of the file, while signal processing involves adding audio effects. Homework: record the output of a synthesizer and program it to have a filter that opens slowly over time. Add a distortion effect and automate it to start in the off position and have it ramp up towards the end of the file. Edit the beginning and ending, and then export the final file. Once again, it is then a straightforward import into your instrument.

3. In this example, the final export should be a collection of different sounds that are combined into a single project. The original elements are prepared separately, the same way as in number two above. After each is completed, import them all into a new project and edit them together. When you play it back, it will sound like a sequence of different elements and may not make musical sense, but that is perfectly okay for use in a wavetable. Homework: create a patch with 15 to 20 different individual sounds. Keep them short, even as short as a single wave form.

4. This final example recycles sounds made in your wavetable synthesizer for reimport into a new wavetable. The idea is to perform a patch using modulators to alter the sound and then record the output into a new file, which is subsequently imported as the source for a new wavetable. Homework: pick an initial wavetable and decide how you want to modulate it. I recommend a real-time modulator such as an X-Y pad or modulation wheel so that you can create a very customized end result. Perhaps your instrument can handle the recording and reimporting of the sound, but if not, then you'll use your DAW for this task. Create a performance and then see what you can do with it after importing the resulting file back in.

Wavetable synthesis uses single cycle waveforms and each of the above examples contains far more than a single cycle, which means that the wavetable synth has to analyze the audio and prepare the file for proper playback. This shouldn't change the way you prepare files except with number three, which involves different sounds and should be edited as single waveform elements.

Figure 5.6 Typical envelope example.

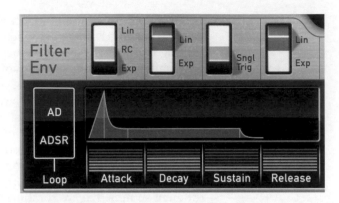

Envelopes

As with most types of synthesizers, the envelopes play an integral part of the sound generation process. Perhaps, with wavetable synthesis, the envelopes are even more important to master because of the way you can use them to control the amplitude and filter, while at the same time controlling the travel of the wavetable. Aligning these together can create an impressive effect that controls the volume up and then back down in sync with the motion of the wavetable. The way to create this is to set the wavetable modulation source to the same envelope that is controlling the parameter you want to match. The end result of an envelope match with the amplitude envelope is that as the initial volume rises, the speed of the wavetable travel also increases. Then, as the release returns to a zero level, the speed of the wavetable travel slows down.

Double Modulation

Many wavetable synthesizers have a modulation matrix because of how important modulation is to this type of synthesis. The benefit to this is that you can create complex modulation chains that are sonically very interesting.

1. Assign an LFO to control the speed of a second LFO.
2. Assign the second LFO to control the position of playback in the wavetable.
3. Assign the modulation wheel to control the speed of the first LFO.

Figure 5.7 Modulation example.

This modulation chain creates an effect where the motion of the wavetable speeds up and slows down as a result of the compound speeds of both LFOs. The modulation wheel connection allows you to have real-time control over the speed for further adjustment.

Waldorf Nave

Waldorf didn't invent wavetable synthesis, but when PPG closed, they took over and even hired some of the PPG team. Waldorf continues to develop new synthesizers using wavetable principles, and have created one of the most important and powerful iOS apps for the iPad. Because of the uniqueness of this instrument and the platform it was developed for, the following is a full profile of its capabilities with the goal of continuing the discussion of how to use wavetable synthesizers.

Figure 5.8 Waldorf Nave app load screen.

Basic Interface

There are five primary pages that encompass everything Nave has available, and each is accessible by touching a menu strip in the top left corner. One of the things that is different about this instrument is that it is designed for a multitouch interface, and so there are high expectations for the touch experience. While the buttons are small, they are easy to use and it is hard to miss the choice you want. It is the touch experience, however, that is one of the most exciting parts of this instrument.

Wave Section

This initial section is where the wavetable is accessible, along with spectrum controls, the mixer, oscillator sections, and a keyboard. In the middle of the screen, there is a graphical representation of the wavetable, with full colors and the ability to touch and drag the graphics around to explore the sound in three dimensions. The representation is very informative about the harmonic content and is useful when performing

Figure 5.9 Nave interface.

because you can see a direct representation of the content you are performing.

The wavetable representation view is customizable with wave view options vs. spectrum view. It is possible to zoom in on the waveform and also adjust the color scheme, all using buttons immediately below the display. In addition to the initial section with the graphic representation, you can also enter a full screen mode that gives you additional tools and functions. These include a ribbon for auditioning sounds, an import file menu, and perhaps the most important tool in the entire app: the edit tool.

The edit tool gives you full control over a large number of parameters that affect the wavetable. These include:

Level
Expand/Contract
Permute
Rotate Waves
Shift Waves
Rotate Partials
Shift Partials
Gyrate
Random

Figure 5.10 Nave edit window.

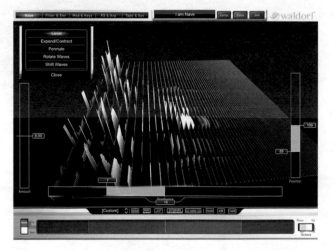

The sliders adjust the portion of audio that is manipulated, leaving portions outside of the selection unchanged. The amount of mangling that can occur in this editor is impressive, along with the amount of fine tuning and smoothing out. This editor is clearly the result of a team that has years of experience working with wavetables and they have poured that experience into an amazing interface that looks and works great.

Oscillators and Mixer

The other welcome addition to Nave is an oscillator section with traditional, simple waveforms. These can be used on their own or to enhance a wavetable instrument. The mixer is a good example of a smart-touch interface because it fits into a very small footprint on the screen, but when

Figure 5.11 Nave mixer in action.

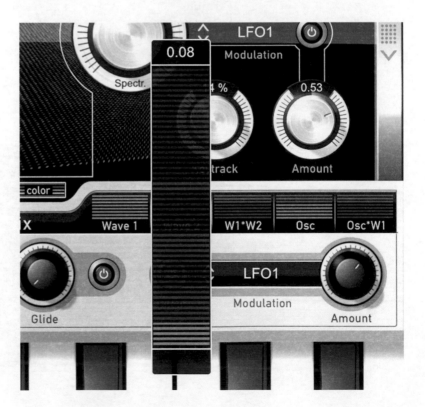

you touch one of the options, then it grows to a significantly larger fader with higher resolution data steps. They even threw in a ring modulator between W1 and W2, and then the Osc and W1.

Waves

There are two wavetables available and you can switch between them by swiping up and down along the left edge. The layout is clear about signal flow, and makes it easy to adjust modulation sources and amounts. As you adjust the starting point for wavetable playback, a red line moves across the 3D rendering. Unfortunately, this isn't animated during playback, but this is likely due to limited resources on a tablet device.

Parameters such as travel are compounded with modulators in the matrix, so if you are using both, then it is important to keep track of what is being modulated. The spectrum section is also able to receive modulation data and this can vastly change the sound of the instrument at extreme settings.

Überwave

The Überwave option is used to double up on several of the oscillator waveforms, much as the unison setting does on many other synths. The density knob adds up eight total duplicate voices, while the spread option detunes each of them. These are great sounding oscillators and they can be modulated, filtered, and processed in Nave just as you would on any subtractive synth.

Figure 5.12 Überwave controls.

Filter and Envelope Section

The rest of the sections are covered here in less detail because many of them have fairly traditional parameters, so they are covered in other chapters of this book. It is the less common items, however, that are the focus of the next paragraphs. In the filter and envelope section, there are three envelopes, one filter, and a drive section. While perhaps it is a little uncommon to have three envelopes, it's certainly not unheard of, and even the linear and exponential switches aren't completely foreign, but are certainly less common.

It is the envelope looping function which is both interesting and highly useful for synthesis. Looping the attack and decay (AD) parameters for the amplitude envelope is something that you may have accomplished previously using an LFO, but this is different because you have access to a greater range of options by looping the envelope. One warning: be careful setting the release to its highest setting because there isn't a panic button to stop it and it will continue on regardless of most attempts to stop it.

The drive effect is also very nice, but not revolutionary. It is always helpful to have a processing option built into the instrument itself for situations when performing live or when you are too lazy to pull up an even better distortion processor. As with all of the other tools in Nave, this distortion module is thought out and useful, while still being incredibly streamlined and simple. It would be nice to have this module as a destination for modulation sources, instead of keeping it separated from everything else.

Figure 5.13 Nave filters and envelopes.

Modulation and Keys Section

The section includes the LFOs, the modulation matrix, the X-Y pad, and customization controls for the keyboard portion of Nave. With ten matrix connections available, both LFOs being full featured and the X-Y pad being pretty awesome, this is a section that you should be visiting fairly often.

LFOs

The LFOs have the standard waveform offerings, with the addition of a very useful sample and hold option. As discussed in previous chapters, sample and hold is often used to sample noise at a predetermined rate to achieve a random output signal. This is useful for creating random pitch or filter sequences in patches, but is also excellent at controlling the wavetable mix levels.

The delay setting fades in the LFO over time, which creates a slow pitch ramp when the LFO is set to control the pitch. The phase parameter adjusts the start of the LFO so that it either starts exactly at zero when triggered, or at any point in the waveform. At a full setting, the phase option lets the LFO run free regardless of when notes are triggered. This means that when you play a note, the LFO might already be halfway through its cycle. The sync options enable the LFOs to work in tandem or reference the project grid, which is useful for connecting through other apps (see Figure 5.18).

Figure 5.14 Nave modulation and keys.

Modulation

The modulation options are pretty straightforward and flexible. When using the matrix, follow the listed steps to establish modulation connections.

1. Assign the modulation source.
2. Assign the modulation destination.
3. Press the power button.
4. Adjust the modulation level.

Keys

There are three different control options and it is highly recommended that you explore these to determine which one you prefer the most. These don't affect the controls of an external MIDI controller, but they can potentially enrich your Nave experience.

1. Traditional Keyboard—This option is the default and looks just like a piano keyboard. The scroll mode is switchable so that dragging a finger along the keys move the keys for different ranges. It can be set to glissando so that dragging a finger plays the different notes. Changing ranges is limited to the Mod & Keys page, and X-Y Touch, which provides modulation control when fingers are moved across the keys.
2. The Blades—This is an alternate control method that is loosely based off of Don Buchla's synthesizer designs. By leaving the tradition piano format behind, there is added flexibility in assigning notes to blades, and developers added chord and scale functionality.
3. X-Y Pads—This is a fun way to control Nave because it maps multiple parameters onto separate pads that are easily

Figure 5.15 The blades.

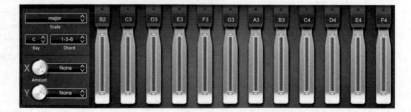

accessible for playing using thumbs. There is a limitation because you can't trigger notes using them, which means that you either have to use them in conjunction with an external controller or use the arpeggiator in hold mode.

FX and Arp Section

This section contains excellent effects for sculpting the ultimate patch. Here is a list of available effects:

- Phaser
- Flanger
- Chorus
- Stereo Delay
- Reverb
- Parametric EQ
- Compressor

In addition to these effects, there is also a very capable arpeggiator. One of the most enjoyable ways to test out modulators is to turn on the arpeggiator in hold mode, and trigger a pattern to start playing. After it starts, you can move to other sections of Nave and make changes while still hearing the original pattern as it responds to the new settings.

Tape and Sys Section

This section houses the master controls and a four-track recorder. The biggest part of this section is the recorder,

Figure 5.16 Nave tape and system.

which is designed to look like a tape machine. There really is no reason for the tape machine theme, but it is pretty cool looking and it works adequately well. The biggest limitation is that the recorder only records audio and not MIDI, which means that whatever you record can't be edited nearly as easily as MIDI. The biggest strength of this system is that you can record the output of sounds and immediately use them as the source of a new Nave instrument.

An even better option is to use Inter-App Audio and record your Nave performance directly into an app like Garageband, which is capable of more than four tracks and has enhanced editing and effects. Inter-App Audio has to be developed into both the master and slave apps, and, when connected, the tempos and transport are locked together. When you press record in either app, the master app records the audio into one of its tracks. This functionality

Figure 5.17 Inter-App Audio in Garageband.

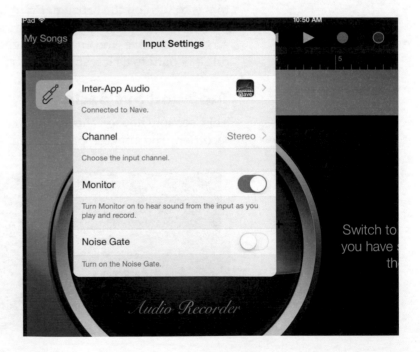

is immediately available with any apps that are Inter-App Audio compatible.

Nave is also WIST compatible, which is another inter-app connection protocol, but is less used now that iOS has Inter-App Audio functionality. WIST was created by Korg and is useful for syncing multiple instruments together over a wireless Bluetooth connection. Transport and tempo information are transmitted and make it possible for both devices to play together.

MIDI

In this section, there is also a MIDI device menu and a MIDI map menu. The device menu shows all currently connected devices, including wireless connections, which is useful for troubleshooting and establishing the MIDI clock. The mapping menu shows all currently mapped parameters and the incoming Continuous Controller Messages (CC) that control them. Most parameters can be mapped to MIDI, even when they are not in this menu. To add a new map connection, double click on any parameter in Nave and follow the instructions that pop up. The double-click option is actually a little annoying because, in many apps, double clicking on a parameter resets it to the default value and, in Nave, there isn't an equivalent function.

Figure 5.18 Nave MIDI mapping.

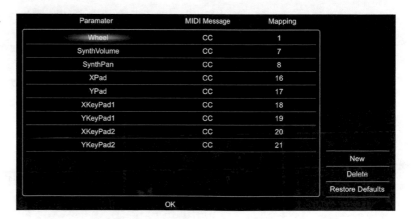

Paramater	MIDI Message	Mapping
Wheel	CC	1
SynthVolume	CC	7
SynthPan	CC	8
XPad	CC	16
YPad	CC	17
XKeyPad1	CC	18
YKeyPad1	CC	19
XKeyPad2	CC	20
YKeyPad2	CC	21

New

Delete

Restore Defaults

OK

Nave Summary

Nave is an incredible app that brings serious sound design and synthesis to a powerful, mobile environment. It would certainly be tempting to buy an iPad just to have this instrument, and is a clear reason to not buy any other tablets. Very few other wavetable synthesizers provide as much power and flexibility as Nave: it demonstrates what is possible when a wavetable synthesizer is designed for a multitouch experience.

Other Wavetable Instruments

Wavetable synthesis is going through somewhat of a renaissance of late, with a new instrument being released from Waves, wavetable synthesis incorporated into several Nord instruments, Waldorf continuing to develop new instruments, and even MachFive incorporating it inside a sampler.

The power of wavetable synthesis is that it doesn't necessarily have its own sound and can sound like a lot of things depending on the waves in the wavetable. Acoustic instruments are perhaps the hardest to emulate using wavetables, but there are plenty of other options for doing that.

Example Projects

Explaining projects for wavetable synthesis is tricky since every instrument is different and has different waveforms. To keep things manageable, the waveforms associated with each wavetable are either readily available online or provided as downloads at the official text website. You are also always welcome to use your own recordings or files.

Recipe 1: Voice Patch

A favorite wavetable source is spoken or synthesized speech, and instruments like Nave actually include a built-in speech synthesizer. If the file is played at a normal rate without severe modulation, then the words would remain recognizable.

Figure 5.19 Recipe 1 – Voice Wavetable in Nave App.

When played as a single waveform at a time and then sometimes scanning through to other portions, you can achieve a metallic or industrial type of sound that sometimes still has elements of speech. There are very few other ways to achieve this sound and using the X-Y pads live often has very pleasing results.

Recipe 2: Hulusi Gourd

Reed instruments are excellent sources for wavetables because they sound good when split into single cycles; and, also, as the playback location is modulated, they also have a very interesting sound. The *hulusi* is a tradition Chinese instrument that has a unique, but familiar sound and is easy to manipulate. The focus of this patch is in setting the LFO modulators to control the wavetable performance.

Recipe 3: Arpeggiated Sawtooth

It might seem that manipulating a simple sawtooth with an associating arpeggiator is better accomplished using the original synth and arpeggiator, but if you export an audio of the pattern, then there are some things you can accomplish as a wavetable that greatly alter the original to make an entirely new sound. The focus of this recipe is in modifying and modulating the travel parameter to enhance the arpeggiation and even throw in a stutter effect.

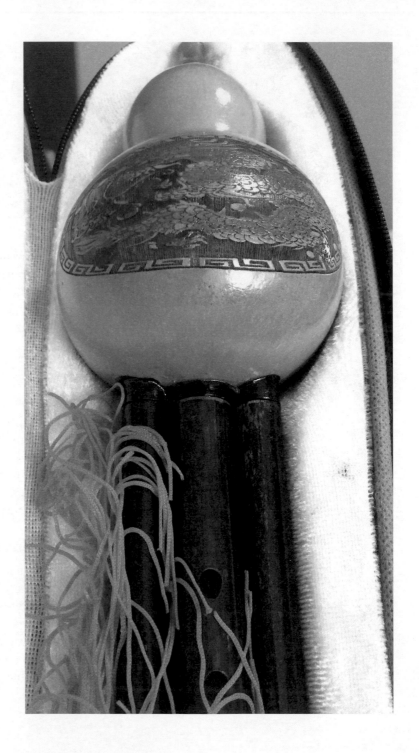

Figure 5.20 Recipe 2 – The Hulusi is a popular traditional Chinese instrument and is sampled for this Wavetable.

Figure 5.21 Recipe 3 – Using Nave's arpeggiator to create a new Wavetable.

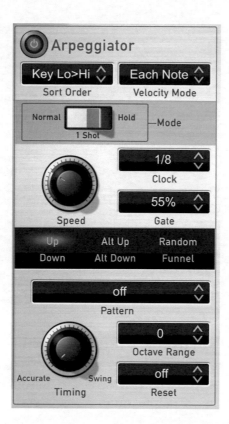

Recipe 4: Simple Waveform Collection

This recipe uses a wavetable made of a collection of single-cycle waveforms taken from all of the primary types. The individual waves are not tied together, so it won't make as much sense to travel through them. Though, there are some alternate uses of wavetables that don't require as much motion through them. The sound sources for the wavetable are all recorded from vintage synths and the synth is used to mix and match their sounds, while providing access to envelopes and filters. The wavetable can still be swept through, but don't be surprised if you don't like how it sounds. On the other hand, it may be the exact sound you are looking for.

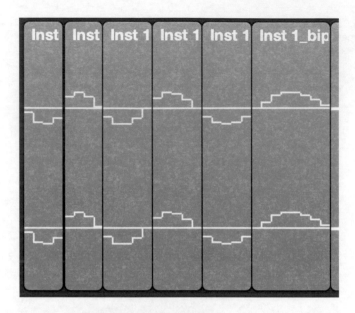

Figure 5.22 Recipe 4 – Cutting single waveforms in Logic Pro X to create a new Wavetable.

Recipe 5: Heavy Metal

This final recipe for the wavetable is based on sounds sampled from metal objects. The sounds were edited and mixed in a separate audio editor and then exported as a single file.

Figure 5.23 Recipe 5 – Sampling a household item as a source for a Wavetable.

The uniqueness of the source material mixed with the exciting modulation of wavetable synthesis creates a sound you won't soon forget . . . it might haunt your dreams. The focus of this recipe is to accomplish some seriously intense audible madness.

Wavetable Synthesis Summary

Wavetable synthesis is one of the most interesting types of synthesis because it is capable of so many different sounds and sonic results. It can focus in on single cycles, or speed up a file until it modulates into a completely different sound. There are applications for this in almost every genre of music and in both the studio and live performance. It is one of the few synthesis types that by default has to live in the digital world, and it makes good use of the functionality that lives in the digital realm.

GRANULAR SYNTHESIS 6

Out of all the synthesis types covered in this book, granular synthesis is the newest to be implemented into commercial synthesizers. This is not to say that granular synthesis theory is new; on the contrary, granular synthesis theory has its roots dating back to the sixteenth century. Implementing granular synthesis, on the other hand, is very new indeed. Granular synthesis had not been properly and commercially introduced until the last decade or so. Granular synthesis is often shrouded in mystery and, many times, granular synthesists are not even completely sure how it works. This synthesis type is often explained in vague terms and definitions. This chapter will serve as a means of demystifying this esoteric synthesis technique in hopes to convey its awesome sound-generation capabilities.

What Is Granular Synthesis?

In its most basic form, granular synthesis is the means of creating a single sound through a variety of extremely short snippets of sampled audio called grains. Picture a roaring crowd at a sports game. The sound emitting from the crowd is heard as a roar, but it is in fact made up of individual dialog and shouting happening all at once. In the same vein, think of a digital picture. Although a digital picture is interpreted as a single picture, it is in fact made up of a number of small, colored dots known as pixels. The fact that we interpret a bunch of incoherent dialog and shouting as a single roaring sound or a variety of colored dots as a single picture is the core concept behind granular synthesis. Like with crowds or digital pictures, each grain in a granular synthesizer is unintelligible by itself, but it is when it is combined with thousands of other grains that it becomes a coherent, unified sound.

Because of the sheer number of grains and controls that are necessary in order to create desired sounds using granular synthesis, it has only become viable as a synthesis format with the advent of computers and digital signal generation. More accurately, it was not until computing power reached a high enough point, about a decade ago, that granular synthesis could truly be adopted. With computing power being increased almost daily, granular synthesis will hopefully become adopted more and more by sound designers and composers in the near future.

Additive Synthesis vs. Granular Synthesis

If reading the brief definition of granular synthesis above sparked your memory of additive synthesis (covered in Chapter 4) you're not alone. In fact, many people actually confuse the two forms of synthesis. It is easy to see the similarities of building a single sound with a large number of individual parts. With additive synthesis, we use a number of individual sine waves to create a single sound, while with granular synthesis, we use a number of individual grains to do the same thing. So how is granular synthesis different from additive synthesis? The simple answer is that additive synthesis uses pure sine waves that are added together to create a tone, while granular synthesis uses a flurry sampled sounds to create a tone.

The Grain

We have made repeated mention of grains in reference to granular synthesis. So what exactly is a grain? In short, a grain is a single sound source with a designated duration that will be combined with many other grains in order to make a final sound using granular synthesis. A grain is similar in theory to a harmonic or partial used in additive synthesis, but the grain is typically much more complex.

The grain is the fundamental building block of granular synthesis. Grains can be made up of any sound source imaginable.

Figure 6.1 The grain and its contents.

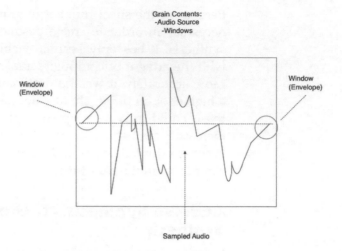

Figure 6.1 The grain and its contents.

In a similar fashion to additive synthesis, granular grains can contain a single sine wave. However, the beauty of granular synthesis is that it is not limited to sine waves, though such is the case with additive synthesis. Besides sine waves, a grain can be made up of any simple or complex waveform. Sampled sounds are most often used for creating grains. This means that any sound, be it a sampled voice, piano, explosion, or siren, can make up a grain. When creating a grain, there are two ways to go about it. The first way is to use short snippets of sound that are the same duration of the individual grain. By doing this, the full length of the sample or synthesized tone will be heard each time the grain sounds. The second way involves using longer starting tones in order to create a number of grains, with each grain being made up of a different section of the sound. Although these two methods may seem similar, they will produce very different results.

Another powerful attribute to grains is that they can be created using sounds produced by any other forms of synthesis such as FM or wavetable. Being able to create a grain from any sound source desired is an extremely beneficial function. Grains can also be created using white or any other color noise. It is important to note that, by themselves, each grain is rather insignificant and is oftentimes perceived as more of

Figure 6.2 Two approaches to using sampled audio for grain content.

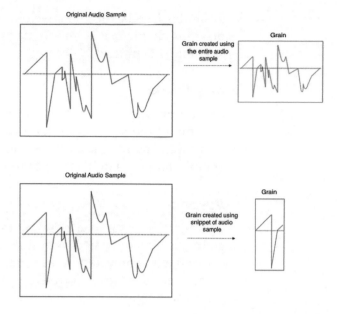

a transient sound or "whoosh" when listened to independently. As we will see later, it is when grains are combined into what are known as granular populations that interesting and desired sounds can be created.

Grain Duration

Once a sound source is decided upon for each grain, the user must set a duration for the grain. Typically grains are set to a duration between one and 50ms (milliseconds). As we learned earlier, in order to interpret pitch, a sound must be played for a minimum duration. Although we will not go into the various mathematical formulas used to determine these durations, it is important to know that typically, a minimum length of 13ms is required for higher frequencies while a minimum of 45ms is required for lower frequencies.

The duration of the grain is determined with an envelope generator much like the ones used in the previous synthesis formats. The term *windowing* is often used for adding a duration envelope onto individual grains. Grain envelopes have a few parameters that call for adjusting and, although they act in similar fashions to the envelope parameters discussed in previous chapters, it is important to discuss them in regards to granular synthesis.

Amplitude

The amplitude parameter of a grain envelope generator determines the max amplitude that the grain will sound at. This control allows the user to designate how loud each grain or granular population is in relation to the other grains.

Duration

The duration parameter determines the full duration of the grain. As already stated, a grain can be anywhere between one and a few hundred milliseconds in length. The duration parameter of the grain's envelope generator is what is used to set this length.

Attack

Much like the envelope generators of other synthesis formats, the attack time of a grain envelope determines the amount of time it takes for the grain to reach full amplitude. The attack time is extremely important in regards to granular synthesis because it shapes the sound itself. When a sound makes a sudden jump from 0dB to anything audible, a resulting distorted transient will be heard that sounds much like a click or pop. Because granular synthesis utilizes thousands of tiny audio fragments, the risk of creating clicks and pops is very great. Therefore the attack parameter of the grain's envelope generator is used in order to smooth out each grain as they are heard.

Sustain

The sustain parameter of a grain's envelope generator determines the amount of time the envelope remains at its maximum amplitude. At first glance, the sustain parameter may seem to be performing the exact same function as the duration parameter. Although they both ultimately affect the length of the grain heard, they are two different controls and special care must be taken in order to ensure they are not fighting against each other.

Decay

The decay parameter of a grain's envelope generator determines the amount of time it takes for the sound to fall back to an inaudible level. Like with the attack parameter, the decay parameter is crucial for eliminating audible clicks and pops when the sound falls back down to 0dB.

A Few Forms of Granular Synthesis

Although granular synthesis is a single-sound generation format, there are at least two different ways of achieving granular synthesis. These two formats are known as synchronous and asynchronous granular synthesis. In synchronous granular synthesis, grains are added to the overall sound texture at exact repeating intervals, or periods. In asynchronous granular synthesis, the rate of the grains is randomized. Synchronous granular synthesis is more easily controllable and is therefore used often to create the exact, desired sounds. Because of the randomization inherent with asynchronous granular synthesis, it is most often employed to create rich, evolving, interesting textures that might not have been originally intended.

Although these are the two main forms of granular synthesis, a third form known as quasi-synchronous granular synthesis may be employed. This third form is really just a median between the two forms and allows the user some randomness without going full asynchronous.

Whether using synchronous or asynchronous granular synthesis, a user will typically have a few functions they can control in order to craft the sound. Although the functions will be the same in both synchronous and asynchronous granular synthesis, the effect they will have will be slightly different.

Pitch

The first and most important parameter that will need to be determined is the pitch of the sound. Typically, pitch is determined by the various sample rates grains are played back at. For example, a sound sampled at 48kHz played back on a granular synthesizer at 44.1kHz will be lower in pitch. Likewise, a sample recoded at 48kHz played back at 96kHz will be higher in pitch. The keys on a keyboard will correspond with different sample rates in order to spread different pitches across the keyboard. Although traditional, Western, 12-tone increments are most often used, microtonal tunings are also possible when using granular synthesis.

Grain Density

The grain density control is what determines the amount of grains spread in the sound spectrum at a given point of time. In other words, grain density determines the amount of grains that will be heard per second. The grain density control is one of the more dramatic timbre-shaping controls in all of granular synthesis. When using synchronous granular synthesis, grain density will affect the overall timbre, amplitude, and pitch of the sound. As the grain density increases, the complexity of the sound will increase. The amplitude of the sound will also increase with greater grain density. Rhythmic effects can also be created simply by lowering the grain density to around 20 to 30 grains per second.

Grain density will yield slightly different results when using asynchronous granular synthesis. Because asynchronous granular synthesis spreads grains across the spectrum in a random-like manner, much more complex and unpredictable sounds can be created. Pitch, for example, will not change with grain density when using asynchronous granular synthesis. If one were to set the grain density to, say, 120 grains per second, then 120 grains would randomly be placed in the texture in that second rather than them playing in a repetitive manner. Therefore, pitch will not be related to the grain density feature. Like with synchronous granular synthesis, rhythmic effects can be obtained by lowering the grain density when using asynchronous granular synthesis.

Putting It All Together

So now that we have discussed what makes up a grain, let's examine how to control the sound as a whole. Based off the quantum theory of sound discussed earlier in the chapter, we can see that we do not interpret the sounds created using granular synthesis as individual sound samples played in succession, but rather as a complete cohesive sound. This is mainly due to the short durations of the grains, as well as how they are placed into the texture.

Grain Order

An important thing to note when using granular synthesis is the order of the individual grains. When creating sounds, the order that the grains are placed will make a considerable difference over the sound as a whole. One method when using sound samples is to have the grains organized in an order that matches the sample. For instance, if using a long sample with many grains, each grain containing a small splice of the sample, the grains must be placed in an order that would allow them to reproduce the full sound sample accurately. This means that grains consisting of splices from the beginning of the sample would be placed before grains consisting of splices from later in the samples and so on. Maintaining a grain order that matches the original sound sample, however, is not mandatory and, in fact, many extremely innovative and random sounds can be achieved by not using the exact order.

Grain Stream

The discussion thus far has been concentrated on single grains and groups of grains known as grain populations. Using a single grain population, however, can yield a very basic and uninteresting sound. Because of this, it is necessary

Figure 6.3 Grains consisting of segments of sampled audio can be arranged in their original order or in a random manner.

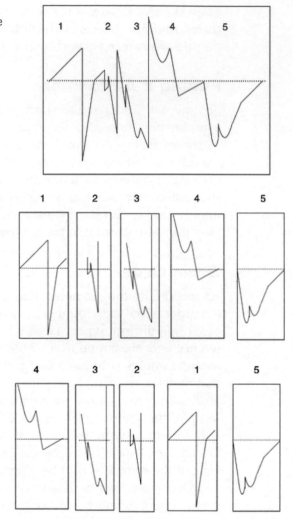

to incorporate multiple grain populations together into what are known as streams. In essence, a grain stream can be likened to a guitar part being doubled or tripled on a recording. With each pass of the same guitar part, new elements will be added to the sound as a whole, creating one comprehensive, interesting part.

The individual grain populations in each stream can be made with various means, but most typically, each sound sample will be used for each corresponding stream, meaning that the first stream will contain grains obtained through a single sound sample while the second stream will contain grains made using a second sound sample. It is important to note that each granular synthesizer may have a different naming scheme for grain streams and populations and also might have different ways of controlling grain populations and streams. We'll take a look later on in the chapter at some of the most popular granular synthesizers available and look at their functions and naming schemes in depth.

Grain Playback Speed

Once grains are determined and created, the speed at which these grains are played back will make a significant difference in the overall timbre of the sound. At slower playback speeds, granular synthesis is capable of producing some of the eeriest, unique, and interesting soundscapes. At faster playback speeds, more novel notes and synth parts begin to take shape until turning into utter chaos and distortion at the highest of playback rates. Adjusting the playback speed of the grains, grain populations, or grain streams is one of the most effective ways of changing timbre using granular synthesis.

Problems Inherent in Granular Synthesis

It can be seen based off of the information covered thus far that granular synthesis requires a lot of fine tuning to say the least. When considering the number of controls that must be determined for each grain, and then multiplying that number by the number of grains that will be used, the sheer number of controls becomes daunting. For this reason, many granular synthesizers will feature complex algorithms to do much of this work for the user.

Borrowed Synthesis

As is the case with most synthesis formats these days, modern granular synthesizers will typically offer parameters that should be familiar to the user, such as filters, LFOs, and envelope generators. Although using these types of controls is not technically part of the granular synthesis process, they are included to aid sound generation. These other synthesis controls will work much in the same way as they do with other synthesis formats in order to achieve desired, predictable results.

A Note on Formant Synthesis

Although not directly related to granular synthesis, a brief understanding of formant synthesis is required in order to utilize the full potential of many granular synthesizers. Formant synthesis is based around the science of formants, or spectral peaks of the acoustic instruments. Formant synthesis is most closely related to physical modeling synthesis, but is its own synthesis format none the less. Formant synthesis is most commonly used to create vocal-like sounds akin to speech. Formant synthesis employs a series of complex band pass filters in order to mimic naturally occurring formants. The type of filters used in formant synthesis, known as formant filters, will often be employed on granular synthesizers. Because granular synthesizers almost exclusively exist in the software realm, formant filters are an easy addition to a granular synthesizer in order to maximize sound-creation potential. It is important to note prior to delving into formant filters that they are not a granular synthesis parameter but, instead, a parameter found on certain granular synthesizers.

Formant Filter

Formants are, in essence, resonances in the human voice that designate aspects of timbre. A formant filter utilizes a series of narrow, band pass filters in an attempt to mimic these resonances. A formant filter will typically allow a user to morph

Figure 6.4 Grendel formant filter.

the sound between a series of set formants with various means of control, such as LFOs and envelope generators. Formant filters can be extremely beneficial on granular synthesizers because they add a depth to the sound creation process that cannot be reproduced through any other means.

The Complex Granular Synthesis Environment

Granular synthesis is most usually used as an all-encompassing term for a variety of similar synthesis formats. Any synthesis format that utilizes sampled audio in order to create short snippets of sound, and then plays them back at varying speeds and frequencies while layering other snippets of audio, is typically referred to as granular synthesis.

However, there are a few other forms of synthesis that use similar principles for sound creation, but are not technically granular synthesis.

Wavelet Synthesis

Wavelet synthesis is often confused with granular synthesis. Although the two synthesis formats are extremely similar, they are in fact two separate synthesis formats and merit a discussion. Like granular synthesis, wavelet synthesis utilizes individual grains that, when combined, create sound. Unlike granular synthesis however, wavelet synthesis has strict rules for its grains. In granular synthesis, individual grains can be of any length and playback speed where as in wavelet synthesis, grain duration and speed are directly related to the pitch of the grain. Therefore, wavelet synthesis offers less flexibility and customization than granular synthesis. Often thought of as yet another synthesis format all together, grainlet synthesis (which is talked about much less often) is, in actuality, just another term for wavelet synthesis.

Glisson Synthesis

Out of all of the synthesis formats confused with granular synthesis, glisson synthesis is perhaps the most similar. Glisson synthesis is almost identical to granular synthesis in that it consists of individual grains, each derived from sampled audio, and played back at various speeds and with different combinations. The main and only difference between glisson synthesis and granular synthesis is that glisson synthesis modifies its individual grains with the use of glissando or glide type effects.

Pulsar Synthesis

Pulsar synthesis is extremely similar to granular synthesis in that thousands of individual sounds are combined together in order to create an end sound. Because of this similarity, it is understandable that it is often confused with granular synthesis. In actuality, pulsar synthesis is considered a form or particle synthesis. Pulsar synthesis utilizes impulse generators in order to create its individual sounds, known as pulsars. Besides using impulse generators to create its starting sounds, pulsar synthesis is nearly identical to granular synthesis.

Sample-Based Synthesis vs. Granular Synthesis

Up to this point, we have constantly mentioned the use of sampled audio in regard to the content of grains. So what then is the difference between sample-based synthesis and granular synthesis if they both make use of samples and sampling? The answer lies in the way the samples are utilized. In sample-based synthesis, entire sound samples are typically used in order to mimic the sound of a particular instrument, or to be able to modulate, mangle, and mutate a sound of your choosing. In granular synthesis, however, extremely short samples, or extremely small snippets of samples are used. When used in this way, the resulting sound coming out of the synthesizer will most often not sound anything like the sample being used. Samples are used just as a means to create grain content.

While we're on the subject, let's examine some ways in which sampled audio can be used in granular synthesis. The first most common technique is to create a number of grains from a single audio sample. Imagine a pianist striking an octave on the lower register of the piano and letting the sound ring out. The sound will begin extremely rich and with high amplitude when the hammers first strike the strings. Once the hammer strikes the string, the sound will then fade down to the resonating strings until it is finally inaudible.

If we had this sound as a sample and loaded it into a granular synthesizer, we could then chop this sample up into a large number of small, 1–50ms samples. If we simply played back each of these small samples in rapid succession, and in order, we'd get something that sounded similar to our original sample.

However, granular synthesis not only allows us to slow down or speed up the playback speed of our samples, but also allows these new samples to be played back out of order, resulting in extremely interesting sounds. What is more, we can create grain populations and grain streams with these samples, meaning we can combine individual samples while adjusting their playback speeds, resulting in something that sounds nothing like the original piano strike we started with. Here lies the true power of granular synthesis. Truly inspiring sounds can be created using this format that cannot be created otherwise.

Another technique for incorporating samples into granular synthesis is beautifully employed by Daft Punk on the track "Robot Rock/Oh Yeah" on of the album *Alive 2007*. As the song opens, two odd, robotic sounds are heard in succession to each other. As the song progresses, the two sounds seem to morph into faintly comprehendible words as they speed up. Then as the two sounds reach their maximum speed, the two sounds are no longer heard as musical tones, but instead are heard as the spoken words *robot* and *human*. Daft Punk accomplishes this effect through the means of granular synthesis.

In order to create this effect, samples of the words *robot* and *human* were taken from the songs "Robot Rock" and "Human after All," respectively. Then, using granular synthesis, the samples were split into a large number of short samples, or grains. In the case of *robot*, individual samples would be created for each letter resulting in R-O-B-O-T. Once each letter is split into individual grains, the word can be elongated by looping each grain resulting in RRRRRRRRROOOOOOOOOO OOOBBBBBBBBBBBOOOOOOOOOOTTTTTTTT. Each subsequent time the phrase is played, the number of times each grain is looped reduces, resulting in the phrase finally being intelligible as a word instead of a sound. As we will explore in the next section, this same technique is employed in the tempo-shifting technology found in software such as Ableton Live.

Granular Synthesis in Various Technologies

Granular Synthesis technology is not only limited to use in synthesizers. On the contrary, many other devices utilize granular synthesis. Although these other technologies are not strictly related to granular synthesizers, they are used in many facets of music production.

Time Stretching

In the recording studio, being able to shorten or lengthen a segment of audio or change its tempo is an absolute must. In years past, when recording on analog tape, changing a piece of audio's speed was accomplished by physically slowing down or speeding up the magnetic tape. By changing the playback speed of the tape, the pitch of the audio was invariably altered—becoming higher or lower in pitch as the tape was sped up or slowed down respectively.

There was no real way to change the tempo without changing the pitch. In modern digital audio workstations, or DAWs, such as Pro Tools, Logic, or Ableton, this is not the case. Being able to manipulate the tempo of recorded audio while maintaining its relative pitch is not only common, but is heavily used. So how then how is this accomplished? The answer lies in granular synthesis.

Time-stretching plug-ins utilize granular synthesis in order to maintain pitch while warping tempo. Let's examine the time-stretching capabilities in Avid's Pro Tools. When elongating a section of audio, the software will find repeating wave shapes and loop them for a set amount of time. By creating short loops, the software is able to elongate the piece of audio as long as the user desires while maintaining its original pitch.

Likewise, when shortening an audio clip, the software will detect repeating wave shapes and delete them, allowing for

Figure 6.5 Granular synthesis is used in modern, digital, time-stretching plug-ins.

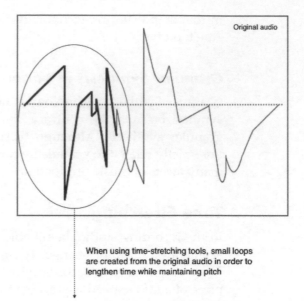

When using time-stretching tools, small loops are created from the original audio in order to lengthen time while maintaining pitch

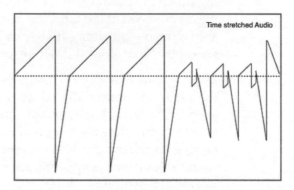

shorter duration while maintaining pitch. The software is effectively using granular synthesis because it is adding or subtracting small snippets of sampled audio. The software will also add envelopes to each piece of audio in order to avoid clicks and pops in the same way that it is done in granular synthesizers. It should be noted, however, that extreme shortening or lengthening of audio will result in sound artifacts, so it is best used in small changes rather than large ones.

Relative Pitch Shifting

Like with time stretching, most digital audio workstations will allow users to alter the pitch of a recorded segment of audio while maintaining the audio's tempo. As mentioned above, pitch and duration are linked together thanks to the laws of physics. Therefore, shifting pitch higher or lower will result in faster and slower audio, respectively. Granular synthesis allows this law to be shattered. Using the same time-stretching technique, modern DAWs will duplicate or delete small snippets of repeating waveforms in order to maintain tempo while altering pitch. When raising pitch, the software will raise the pitch traditionally and then duplicate repeating waveforms in order to bring the new, shorter audio clip back to its original length.

Likewise when lowering pitch, the software will pitch shift the audio down and then delete repeating waveforms on the new, longer audio clip to bring it back to its original duration. Similarly to time stretching, extreme pitch altering will result in audible artifacts, so it is best used in moderation.

Software such as Auto-tune, Melodyne, and Ultra-pitch all utilize granular synthesis in order to perform their respective functions. Had it not been for granular synthesis and its subsequent experiments and implementations, we might not have the time-stretching and pitch-shifting capabilities that we now rely on every day.

Granular Synthesis in Action

Now that we have covered the logistics of granular synthesis, let's take a look at some of the synthesizers available that feature granular synthesis. As mentioned earlier, granular synthesis typically needs a software environment to flourish and, therefore, there are virtually no hardware granular synthesizers, save for a few workstation synths that are capable of performing a small amount of granular synthesis. Because of this, all of the synthesizers we will cover in this section are solely software based.

Camel Audio Alchemy

Alchemy is a powerful, multisynthesis software instrument. Although Alchemy features an impressive array of synthesis formats, the granular synthesis section of Alchemy is where it really shines. Alchemy creates its grains from user-imported or preinstalled .AIFF or .WAV files. Grain length is variable from 2ms to 230ms. Once grains are determined, Alchemy allows users to shape the grain through their *window* feature. In essence, Alchemy's grain window is an envelope and amplitude setting. Once this is determined, users can modify the grain's overall pitch and pan settings.

Alchemy offers users extremely useful time and pitch manipulation tools for further sonic mayhem. Using Alchemy's time-stretching tool, users are able to play back grains at slower and faster speeds, without affecting pitch, by utilizing the technique covered in the time-stretching section of this chapter. One interesting feature Alchemy offers in its time-stretching section is the ability to *freeze* samples at any given point. This is done by extracting a number of grains from a determined section of the sample and looping them.

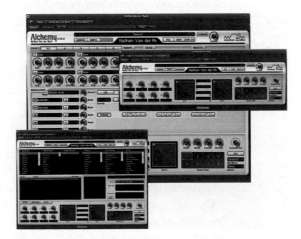

Figure 6.6 Camel Audio Alchemy synthesizer.

Likewise, Alchemy offers users advanced pitch-shifting capabilities. Users can raise or lower pitch without affecting grain playback speed. The ability to alter pitch without affecting grain playback speed has become an absolute must for granular synthesizers and Alchemy does not disappoint in this regard.

Alchemy is a powerful granular synthesizer and has unofficially become the industry's standard granular synthesizer since its inception. Artists like Trent Reznor, Charlie Clouser, and Phaeleh make great use of Alchemy and it can be heard on some of their most influential tracks.

CrusherX-Live!

CrusherX-Live! is a granular synthesizer in the most traditional sense of the word. It is based around the granular synthesis theories of Xenakis, Roads, and Wilkins—three of the early granular pioneers. CrusherX-Live! offers users the ability to determine grain duration and envelopes. One unique feature on the CrusherX-Live! software is its floating buffer, which allows for real-time inputs and outputs. This is especially impressive since many granular synthesizers do not typically offer real-time playability. The designers behind CrusherX-Live! designed the software to not only be used as a standalone synthesizer, but also as a powerful effects unit in the modern DAW environment.

Figure 6.7 CrusherX-Live! synthesizer.

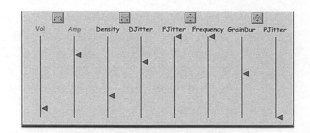

Figure 6.8 Symbolic Sound Kyma X synthesizer.

Symbolic Sound Kyma X

Kyma X is an advanced audio editing and synthesis tool designed for musicians and sound designers alike. When using Kyma X, users are provided with over one thousand synthesis algorithms and synthesis-based effects. Users can audition each preset with onboard sampled audio. Each synthesis algorithm can be fully adjusted and manipulated by the user. Kyma X has been used by sound designers Ben Burtt and Gary Rydstrom in the movies *Wall-E* and *Finding Nemo*, respectively.

Propellerhead Malstrom Graintable Synthesizer

Propellerhead offers a hugely powerful granular synthesizer for their Reason software called the Malstrom Graintable Synthesizer. What is a Graintable, you ask? Well, Propellerhead explains it as a hybrid granular/wavetable synthesizer. In essence, Malstrom allows for traditional granular synthesis while combining the awesome power of wavetable synthesis.

Figure 6.9 Propellerhead Malstrom synthesizer.

By combining these two synthesis formats, extremely complex sounds can be created that must be heard in order to understand their full potential. In addition to its powerful synthesis engine, Malstrom also offers users extreme modulation possibilities as well as most filter types one could wish for.

Image-Line Granulizer

The engineers at Image-Line also offer their own granular synthesizer for their FL Studios software. The Granulizer is a traditional but powerful granular synthesizer. When using Granulizer, users load in their own sound samples in order to attain grains. Although Granulizer is a fairly traditional granular synthesizer, Image-Line offers a few helpful innovations that make Granulizer stand out amongst some of the other granular synthesizers on the market in regards to playability. The two biggest innovations are the inclusion of fully automated and routable grain spacing and wave spacing wheels. The grain spacing wheel controls the spacing of individual

Figure 6.10 Image-Line Granulizer synthesizer.

grains during playback, while the wave spacing wheel controls the amount of grains generated from the sound sample. Having these two controls at your fingertips on a MIDI controller allows for much greater real-time control than is offered on many other granular synthesizers.

Max/MSP

Rather than being a synthesizer in the traditional sense, Max/MSP is a visual programming language environment that is often utilized by musicians. Using the Max software, users can create almost any audio device imaginable, ranging from simple oscillators to signal meters, all the way up to complex, multiengine synthesizers. The Max/MSP software is modular in that users create small pieces of code that serve specific functions that are then connected to other similar pieces of code. Max/MSP can be used to create granular synthesizers. The unique thing about Max/MSP is that a granular synthesizer created inside the software can technically be as complex as the user is willing to create so long as computer power permits.

Figure 6.11 Max/MSP programming environment.

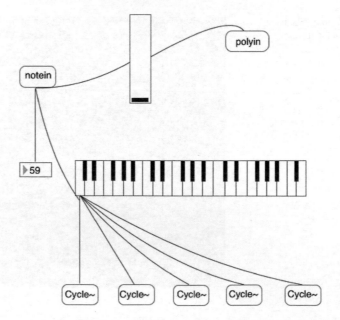

Recipes

Granular synthesis is a fairly new technology. Because of this, it is not met with the same standards that years of research and performing has granted synthesis formats such as subtractive and frequency modulation. Therefore, many of the granular synthesizers available today are completely different from one another in reference to their capabilities. That being said, the parameters discussed throughout this chapter are likely to be found on most granular synthesizers, just perhaps in different configurations. Due to this, we have chosen to create our recipes on the synthesizer Alchemy by Camel Audio. Alchemy's granular environment is one of the best currently offered and features most, if not all, of the parameters mentioned in this chapter. Let's quickly discuss Alchemy's capabilities as well as a brief outline of how to use it before delving into the creation of the recipes.

The Interface

When launching Alchemy, the default screen that appears is the *browser window*. Using the browser window, users can scroll through hundreds of presets in order to gain an understanding of what the synthesizer is capable of. The next screen that can be loaded is known as the *simple window*. Using the

Figure 6.12 The Alchemy advanced window.

simple window, users can have access to a few select parameters as well as performance control functions. The main window to become familiar with, though, is the *advanced window*. The advanced window is the main window to use for sound creation. Each sound source is displayed with each control available: filtering, modulation, envelope generators, and amplifier sections are all visible as well as an in-depth modulation matrix. Morph pads and various performance control features like sequencers, arpeggiators, and effects are also available in this screen. Let's now go "under the hood" and see how to use Alchemy in reference to granular synthesis.

Sound Sources

Alchemy offers four sound sources, *A*, *B*, *C*, and *D*, which can be combined and fully modulated. By default, each sound source is switched to sawtooth waves, and only sound source *A* is engaged. By clicking on black space to the right of the sound source letter, the type of sound can be selected.

When dealing with granular synthesis, we will be utilizing sampled audio. This means that we will select either the "load audio" or "import audio" tab. Using the "import audio" tab, one can load any sample that one wishes for granular manipulation. For our recipes however, we have stuck to the on-board audio files in order for you to follow along if you happen to have access to Alchemy. Once a sample is selected, we can go in and manipulate said sample by clicking the corresponding sound source letter directly above the sound source.

Figure 6.13 Alchemy's sound source view.

Figure 6.14 Sound source expanded view.

Once inside the manipulation section, the "granular" box should be engaged and lit up blue. Once it's engaged, a variety of granular controls become available. These controls include grain volume, grain size, grain density, window, RTime, and RPan.

Grain Volume, Size, and Density

The grain volume, size, and density controls allow the user to manipulate these functions. *Grain volume* adjusts the level of individual grains, *grain size* controls the length of each grain, and the *grain density* controls how the grains are dispersed. Besides grain volume, grain size will yield the most audible results with the sound becoming more distorted and less recognizable, in terms of pitch, when the length is shortened.

Window

The *grain window* control allows users to select the envelope shape that will be placed on each grain. The selections are Hann, Needle, Ramp Down, Ramp Up, Rectangle, Trapezoid, Triangle, Tukey05, and Tukey05e. Each will yield slightly different results with Needle, Ramp Down, and Ramp Up producing the most noticeable.

Figure 6.15 Grain volume, size, and density controls.

Figure 6.16 Window shape selection.

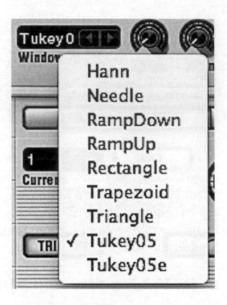

Figure 6.17 RTime and RPan controls.

RTime and RPan

The *RTime* and *RPan* controls allow the user to set random time and pan changes to the individual grains. As one increases the RPan level, individual grains will rhythmically bounce from left to right, causing extremely pleasing effects.

Modulation Matrix

Alchemy features one of the most useful and complex modulation matrixes available. The modulation matrix view has five pull-down menus allowing for five individual modulation sources for each destination. When setting a modulation source and destination, one simply clicks on the desired

Figure 6.18 Alchemy's modulation matrix.

destination which then changes the modulation matrix view showing the five sources corresponding to that particular destination. Modulation sources include:

- LFOs
- AHDSR (Attack, Hold, Decay, Sustain, Release) envelope generators
- Multiple Segment envelope generators
- Sequencers
- A variety of note properties (velocity, key follow, aftertouch, etc.)
- MIDI controller CC messages

Each knob in the advanced screen can be assigned five, separate modulation sources. This is nothing short of amazing. What is more, the numbers of LFOs, envelope generators, sequencers, etc. are endless, meaning each parameter can be modulated by five discreet envelope generators. In essence,

Figure 6.19 Modulation controls.

pitch can be modulated by five independent envelope generators; grain size can be modulated by five more independent envelope generators, etc. The modulation possibilities offered in Alchemy are literally endless.

Once a modulation source is selected, the modulation manipulation section of the interface can be used to customize the source of modulation. This is the section that will be used to set the parameters of envelope generators, set LFO speed, etc.

Master Section

The final section we will use with the granular synthesis engine is the *Master* section. In the *Master* section, we can set things like overall glide, voices of polyphony, main level, and panning, as well as course and fine tuning. Again, like everything else in Alchemy, all parameters of the *Master* section can be modulated by simply clicking a parameter and setting its five, discreet modulation sources.

Now that we have covered the basics of Alchemy's granular synthesis engine, let's dive into the recipes we created for this book and discuss how we created them.

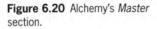

Figure 6.20 Alchemy's *Master* section.

Recipe 1: Cinematic Tension Pad

The first recipe in our list is a slowly evolving cinematic-type pad. During its evolution, dissonant metallic sounds and shimmering midrange synth drones rise and fall. The recipe itself remains interesting and exciting throughout a long time period with only one note held. We began this patch by loading three Alchemy audio files: "Feedback-E1," "MAX-Grain-Ambience-Maj-A2," and "DarkSpace-C2." Inside of the "Feedback-E1" sample, we set the panning a small amount to the left and the grain volume turned all the way up. The grain size was fairly long with the size knob set to around 175ms. Next, the density was set to seven grains each with a trapezoid-shaped window. Finally, RTime was set to around 75% and RPan was completely disabled. Moving onto the "MAX-Grain-Ambience-Maj-A2" sample, grain volume was turned fully up while grain size was set to around 120ms. Grain density was set to one grain, causing it to sound quite distorted with a trapezoid-shaped window. RTime was set to about 48%, while RPAn was set to around 82%. Next, the "DarkSpace-C2" sample featured a full grain volume and about a 112ms grain length. The grain density however, was set to five grains with a rectangular-shaped window. Finally, RTime and RPan were both set to around 48%. Lastly, a single envelope generator was used to control the amplifiers output

Figure 6.21 Screenshots of Cinematic Tension Pad patch.

featuring a slight 0.001-second attack, full sustain, and a 0.37-second release.

Recipe 2: Pleasantly Dissonant Pad

The second recipe on our list is actually a fairly elegant sounding, subtly percussive pad with an evolving dissonant quality to it. It was created using three Alchemy sound samples: "FlightOfAngels-Min-D2," "Dark Emulation," and "DeepSpiral-C1." The "FlightOfAngels-Min-D2" sample featured a fully cranked grain volume and around a 180ms grain length. The grain density was set to four grains, each with a shaped window. Finally, RTime was completely disengaged while RPan was set to around 62%. The "Dark Emulation" sample featured a slightly lower grain volume (right around 1.75dB) with a grain size of around 114ms. Grain density was set to six grains with a Tukey05 window shape as well. Finally, RTime was set to around 42%, while RPan was set to around 40%. The "DeepSpiral-C1" sample featured around a 8.3dB grain volume, which was being modulated by an envelope generator set to have full sustain and release. The grain size was set to around 114ms with a density of five grains. Each grain featured a Tukey05 window shape. Finally, both RTime and RPan were completely disabled. The overall

Figure 6.22 Screenshots of Pleasantly Dissonant Pad patch.

amplifier featured an identical envelope to the third sound source's grain volume envelope in that both sustain and release were set to 100%. Finally, the amplifier envelope featured a 0.001-second attack time.

Recipe 3: Plucked Ambient Thump

The third recipe we created features a heavy pluck sound, which reminds one of the lowest string on a steel string baritone acoustic guitar being strongly plucked with a metal pick. Blurred sounds seem to emanate from the plucked sound almost like it was being fed through a slow, motion delay-type effect. The sound is comprised of four Alchemy samples: "Autoharp-C1," "Violin-Pizz-A2," "FB01-Eqota," and "Clavinet-C0." The "Autoharp-C1" sample is set to full grain volume and a grain size of around 145ms. The grain density was set to seven grains with a trapezoid-shaped window. RTime was set to around 67% and RPan was set to 100%. Moving onto the "Violin-Pizz-A2" sample, grain volume was set to full volume, while grain size was set just short of full at around 216ms. Grain density was set to five grains, each with a Tukey05 window shape. Finally, RTime was completely disengaged, while RPan was almost fully engaged at around 85%. The "FB01-Eqota" sample featured an almost full grain volume set to around 2.75dB with a grain size of around 170ms. The grain density was set to full with ten

Figure 6.23 Screenshots of Plucked Ambient Thump patch.

grains, each with a Tukey05 window shape. RTime was set to around 60%, while RPan was set to around 94.5%. Finally, the "Clavinet-C0" sample featured full grain volume, a grain size of 165ms, full grain density, Tukey05 window shape and RTime and RPan settings of about 46 and 39%, respectively. A sine wave LFO set to 1/4Beat Sync was routed to the RPan level of sound source *D*, Grain volume of sound source *C*, and both RTime and RPan of sound source *B*. Lastly, a basic amplifier envelope with 100% sustain, 0.37-second release, and 0.001-second attack times was set.

Recipe 4: Tyrell Corporation Pad

The fourth recipe was meant to mimic some of the synth sounds found in the favorite movie of one of the authors—*Blade Runner*. The sound is made up of three Alchemy samples: "BubbleWorld-C2," "Smear Atmo C2," and "Trombone-C1." The "BubbleWorld-C2" sample featured a full grain volume and around a 115ms grain size. The grain density was set to five grains with trapezoidal windows. Both RTime and RPan were set to 0%. The "Smear Atmo C2" sample featured full grain volume, size, and density, with a Tukey05 window shape. Finally, RTime and RPan were set to 70 and 90%, respectively. The "Trombone-C1" sample had a full grain volume with a grain size of around 200ms. The grain density is set to eight grains with triangle-shaped windows. Finally, RTime and RPan are set to 36 and 60%, respectively. Lastly,

Figure 6.24 Screenshots of Tyrell Corporation Pad patch.

an envelope generator with full sustain and release parameters with a 0.001-second attack is routed to the synthesizers amplifier.

Recipe 5: Low Burst Pad

The fifth recipe in our list is interesting in that it utilizes a sample of a human laugh to create a low, weird effect that pans from speaker to speaker via LFO modulation. Then, three low brass samples are played to create a sinister low growl. The sound consists of four Alchemy samples: "Tuba-C0," "Trombone (sfz)," "Trombone-C1," and "Jonni-Laugh." The "Tuba-C0" sample featured full grain volume and size with a grain density of nine grains each with a RampDown window shape. Finally the RTime and RPan controls were set at about 32 and 84%, respectively. The "Trombone (sfz)" sample consisted of a full grain volume and a grain size of around 207ms. The grain density was set to eight grains with a trapezoid window shape. RTime and RPan were set to 18 and 72%, respectively. The "Trombone-C1" featured full grain volume, size, and density with a triangle-shaped grain window. The RTime was set to 68%, while the RPan was cranked up to 100%. The final sample, "Jonni-Laugh," featured full grain volume

Figure 6.25 Screenshots of Low Burst Pad patch.

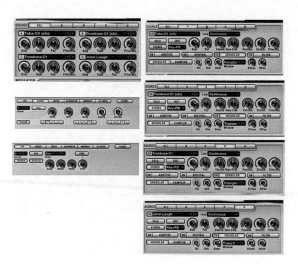

with a grain size of around 189ms. Grain density was set to nine grains with a Tukey05 window shape. Finally, RTime was fully engaged, while RPan was set to around 92%. A sine wave LFO was set to around 0.28Hz and was modulating the pan control on the "Jonni-Laugh" sample. It should be noted that the LFO was set to "TRIGGER," meaning that each time a key was pressed, the LFO would restart its cycle. Finally, an envelope generator with a 0.001 second attack and fully cranked sustain and release parameters was set to module the amplifier and each grain volumes of sound sources *A*, *B*, and *C*.

Recipe 6: Glistening Auto-Harp

Although not utilizing an auto-harp sample, the sixth recipe in our list is reminiscent of the instrument with an added cosmic flair. The sound consists of three Alchemy samples: "014-SantoBrasil," "Sitar (sfz)," and "dw8-bars (sfz)." The "014-SantoBrasil" sample featured full grain volume with a grain size of around 115ms. The grain density was set to four grains with a Tukey05 window shape. Both RTime and RPan were completely disengaged. The "Sitar (sfz)" sample featured full grain volume with a fairly short grain size at

Figure 6.26 Screenshots of Glistening Auto-Harp patch.

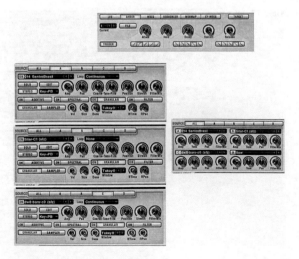

around 16ms. Grain density was set to full with a Tukey05 window shape. Finally, RTime was set to around 68% with RPan completely disengaged. The "dw8-bars (sfz)" sample featured full grain volume, size, and density with a Tukey05 window shape. Both RTime and RPan were set to 100%. A basic envelope generator with a 0.001-second attack, 100% sustain, and full release was routed to the grain volume of the "Sitar (sfz)" sample as well as the overall amplifier.

Recipe 7: Grain Length Modulated Plucking

The seventh recipe is a constantly evolving patch that at first sounds like a plucked Stratocaster, but then, when the LFO kicks in, the entire sound becomes thin and distorted before returning back to the plucked Stratocaster sound. The patch consists of three Alchemy samples: "Steel String- E1," "Ghosts of gaia-Min-E2," and "Appalachian Dulcimer (sfz)." The "Steel String-E1" sample contained a full grain volume with a grain size of about 112ms. The grain density was a full ten

Figure 6.27 Screenshots of Grain Length Modulated Plucking patch.

grains with the Hann window shape. RTime was set to around 82% with RPan fully engaged. The "Ghosts of gaia-Min-E2" sample contained full grain volume with a grain size of around 110ms. Grain density was set at four grains with the Tukey05 window shape. Finally, both RTime and RPan were completely disengaged. The final sample, "Appalachian Dulcimer (sfz)," contained a full grain volume with a grain size of about 110ms. The grain density was identical to sound source *B* at four grains with the Tukey05 window shape. Both RTime and RPan were fully disengaged. A sine wave LFO set to 0.11Hz was routed to the grain size of each of the three sound sources. Note that the "TRIGGER" function of the LFO was not in use, meaning the LFO was free running. Lastly, a basic envelope generator with a 0.001-second attack, 100% sustain, and 0.37-second release was routed to the amplifier.

Recipe 8: Glistening Bells

This recipe was designed to sound like tubular bells with a haunting ethereal quality. The sound consists of three Alchemy sound samples: "Feedback-E1," FinetoothPad-Maj-F3," and "Electric Synth Keys v127." The "Feedback-E1" sample contained full grain volume, size, and density with the Tukey05

Figure 6.28 Screenshots of Glistening Bells patch.

window shape. RTime was set at about 20%, while RPan was set to 100%. The "FinetoothPad-Maj-F3" sample featured a full grain volume with a grain size of about 110ms. The grain density was set to nine grains and featured a Tukey05 window shape. RTime was set to about 77%, while RPan was set to 62%. The final sample, "Electric Synth Keys v127," contained a full grain volume with a grain size of about 122ms. The grain density was set to four grains with a Tukey05 window shape. Both RTime and RPan were turned all the way off. An envelope generator with a 0.0085-second attack, full sustain, and 16-second release was routed to the amplifier as well as the grain volumes of sound sources *A* and *B*.

Recipe 9: Dampened Metallic Hammer Shimmer

The ninth recipe in our list was designed to sound like a mix between steel drums and a celesta with a shimmering electronic quality. The sound consists of three Alchemy samples: "F2M C1 (sfz)," "013 Etiopia," and "Dulcitone-f (sfz)." The "F2M C1 (sfz)" sample contained a full grain volume with a grain size of about 18ms. The grain density was set to ten

Figure 6.29 Screenshots of Dampened Metallic Hammer Shimmer patch.

237

grains with the Tukey05 window shape. RTime was set to about 92% with RPan fully engaged. The "013 Etiopia" sample contained a full grain volume with a grain size of about 111ms. Grain density was set to nine grains with the Tukey05 window shape. RTime was set to about 30%, while RPan was set to about 7.5%. The "Dulcitone-f (sfz) sample featured a grain volume of about 6.98dB with a grain size of 230ms. Grain density was set to eight grains with the Tukey05 window shape. Finally, RTime and RPan were both set to 100%. An envelope generator with a 0.001-second attack, full sustain, and 0.37-second release was routed to the amplifier.

Recipe 10: Rhythmic Modulation

The final recipe we created was designed to be a rhythmic patch that contained both bass and lead sounds playing together while the rhythmic pulses maintained in the background. The patch is comprised of four Alchemy sound samples: "Harp (sfz)," "Dulcitone-f-A2 (sfz)," "Cronofy," and "FracturedSignal." The "Harp (sfz)" sample featured full grain volume with a grain size of about 112 ms. Grain density was set to five grains with a Tukey05 window shape. Both RTime and RPan were turned totally off. The "Dulcitone-f-A2 (sfz)" sample contained full grain volume with a grain

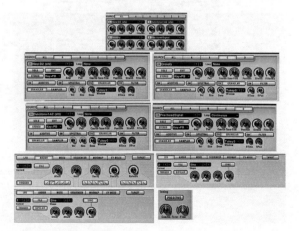

Figure 6.30 Screenshots of Rhythmic Modulation patch.

size of about 150ms. Grain density was set to five grains with a Tukey05 window shape. Both RTime and RPan were fully disengaged. The "Cronofy" sample contained a full grain volume with a grain size of about 150ms. Grain density was set to ten grains with a Tukey05 window shape. RTime was set to about 63% with RPan set at 100%. The final sample, "FracturedSignal," features a full grain volume with a grain size of about 114ms. The grain density is set to four grains with a Tukey05 window shape. Both RTime and RPan are completely disengaged. One sine wave LFO is beat synced at 1/2T and is routed to the grain size of sound sources A, B, and D. A second sine wave LFO set to 0.09Hz is routed to the overall pitch of the synthesizer by +/-2 semitones. Lastly, an envelope generator with a zero-second attack, hold, and decay, with full sustain and release is routed to both the overall amplifier and grain volume of sound source A.

Historical Perspective of Granular Synthesis

Granular synthesis, although new in implementation, is actually quite old. As stated earlier, granular synthesis theory began at the onset of the quantum theory of sound. These early theorizations of granular synthesis, however, only existed in vague theories. It was not until Dennis Gabor theorized a system in 1946 that used grains to produce sound that granular synthesis theory began to take form.[1] Gabor gave a series of lectures in which he examined Fourier's work on sound. Gabor was not attempting to prove that Fourier's theories were wrong. On the contrary, Gabor believed Fourier's theories to be correct, just incomplete. Gabor claimed that Fourier's theory of sound did not allow for common sounds, such as sirens, which have variable pitches. Gabor went on to state that Fourier's theories could only be used to create one sound at a time. Likewise, Gabor was among the first to take criticism with Fourier's use of sine waves due to their infinite nature. Gabor stated that duration of sound was a crucial aspect in defining it.

Gabor went on to conduct a wealth of research on hearing. Most notably, Gabor conducted research involving the time it took for people to interpret sound as having pitch. For example, Gabor found that frequencies between 500 and 1000Hz must have a duration of at least 10ms in order to be interpreted as tone rather than noise.

Gabor went on to build a variety of tone-producing machines that used primitive granular synthesis techniques. One such device was built from a film projector. Gabor ran a film through a projector that would move at constant velocities. As the film moved, it would pass by a rotating drum which contained a number of slits that allowed light to be projected onto a photocell. The photocell would then convert light into sound. After tinkering with the device, Gabor decided to add more slits to the drum. Gabor realized that this would produce tones. However, there would be a loud click at the start of every cycle that was undesirable. By grading the window, the sound would effectively fade in and out, which got rid of the clicks. This can be considered the first use of grain envelopes in granular synthesis. Gabor's research into the perception of sound and hearing paved the way for granular synthesis.

Possibly going off of Gabor's research, the German company Springer produced a device known as the *Tempophon* which was similar to Gabor's device. The Tempophon was installed in a number of early electronic music studios and is often considered the first commercial device that utilized granular synthesis.

The Godfather of Granular Synthesis

Granular synthesis would advance a step further in the 1960s in the hands of Greek composer Iannis Xenakis. Xenakis was born in 1922 in Romania to parents who were both interested in music.[2] Introducing Xenakis to music at an early age helped pave the way for his future career as a composer. In the 1960s, Xenakis came across Gabor's theories and began applying them to a compositional format. Xenakis began composing using grains of sound. Xenakis was perhaps the first to define a grain as containing duration, frequency, and intensity. Xenakis also theorized ways in which these grains could be used in conjunction with one another. This simple idea of combining grains in different ways led to the idea of textures and soundscapes being created using grains. Xenakis went on to record "Analogique B," which has become the first composition to be created using granular synthesis. "Analogique B" utilized hundreds of splices of magnetic tape in order to create its sounds.

After Xenakis, composer Curtis Roads used similar granular synthesis techniques in order to create his pieces of music. Many of the early experiments Roads used in his compositions paved the way to the granular synthesis we know today. The next key player in the inception of granular synthesis was Paul Lansky. In the 1990s, Lansky created a series of compositions known as the *Idle Chatter* series, which utilized granular synthesis in the creation of the pieces. The three songs off of his *More Than Idle Chatter* album that utilize granular synthesis are "Idle Chatter," "just_more_idle_chatter" and "Notjustmoreidlechatter." These three pieces all use hundreds, if not thousands, of small audio fragments of speech that are pitch shifted and placed together in order to create pseudo-melodies. Like Xenakis, Roads, and Lansky, Barry Truax was yet another early adopter of granular synthesis for use in a vast number of his compositions. All these early composers helped solidify granular synthesis's spot as a valid sound creation technique.

Although Xenakis, Roads, Lansky, and Truax all used granular synthesis for their compositions, they used a primitive analog version of what granular synthesis would ultimately become. Early implementations of granular synthesis often took a huge amount of time and resources, and it was somewhat limited in its scope. The advent of computers would finally allow granular synthesis the power needed to be used in full.

As we have seen, granular synthesis uses thousands of individual grains derived from sampled audio. The sheer power needed to not only analyze this sampled audio and chop it up, but also to create individual grain envelopes and playback the grains at various speeds, is extremely large. Therefore, modern computers are necessary in order to fully realize granular synthesis.

As stated before, granular synthesis can still be considered in its infancy. It will continue to evolve as computer power increases. The sound capability that granular synthesis offers is nothing short of amazing. Although it is often shrouded in mystery and not used deeper than the preset level, granular synthesis is capable of some of the most interesting, creative, unique, and awe-inspiring sounds in synthesis. I beg you to try your own hand at granular synthesis and let it open up a whole new world of sonic opportunity.

The Future of Granular Synthesis

Granular synthesis can almost be considered in its infancy. Although it has been implemented and used, it is constantly evolving as computer power reaches higher and higher levels. The next logical step in my eyes for granular synthesis is the use of live audio as grain content rather than sampled audio. Granular synthesizers that allow users to feed live audio into the unit for manipulation is rarely, if ever, seen. What then, is preventing someone from using granular synthesis technology on the fly with a live audio input? The answer is nothing. Although most granular synthesis software mostly work with previously recorded audio, there is really nothing stopping someone from using live audio recordings as the source material for grains, besides developer restrictions. Modern computers have finally reached a level of CPU power that allows live audio to be analyzed, split into grains, and warped without much delay, if any. Therefore, it is my prediction that we will soon see granular synthesizers that allow for live audio sampling in the near future. Some granular synthesis-based effects in fact already allow real-time manipulation of live audio. Native Instrument's Spektral Delay and Smartelectronix's KTGranulator both allow real-time granular synthesis. Many real-time vocal processors and pitch shifters also use aspects of granular synthesis. However, these effects are not fully functional granular synthesizers.

Granular synthesis holds extreme promise solely because it is in its infancy. Other synthesis formats such as subtractive, FM, additive, and modeling have all been around long enough to inherit fixed structures and stringent rules. Although these aforementioned formats are still being implemented and expanded upon, many users expect them to work in certain ways, which leaves little room for developers and engineers to create new and interesting takes on these classic formats. Although I am by no means calling these other synthesis formats inferior (quite the contrary, I love and use each of these formats on almost a daily basis), I am simply stating that granular synthesis holds the greatest potential for future embellishment and ingenuity. Since granular synthesis is still so new in its implementation, preconceived notions about what a granular synthesizer should do have not been fully formed. Developers are free to implement new and interesting functions into granular synthesizers that will hopefully take the technology above and beyond where it currently is.

Notes

1. Timothy Opie, "Sound in a Nutshell: Granular Synthesis," November 12, 1999. Retrieved from http://granularsynthesis.com/hthesis/hthesis.html
2. Iannis Zenakis, "Autobiographical Sketch," 1980. Retrieved from http://www.iannis-xenakis.org/xen/bio/biography.html

VECTOR SYNTHESIS 7

In 1986, Sequential Circuits released their final synthesizer called the Prophet VS, and in doing so, they invented one of the most expressive and powerful performance synthesis types, called Vector Synthesis (VS). The concept is simple with four different sound sources that are mixed together with a joystick on a two-dimensional plane. There are several reasons why VS caught on and is still an important type of synthesis, all of which are related to musicality in performance and efficiency in programming.

The Prophet VS is a digital synth that has a tragic history because it had so much promise, but it came at a time when Sequential Circuits was about to close. It is hard to fix, and parts are rare, which makes it difficult to invest in. However, if you can find one in good condition, then you will certainly enjoy its unique sound. With its extended envelopes and its customized waveforms, you won't be able to treat it like any old synth and will have to learn its nuances.

The key elements in the original VS instrument are four sound sources that can be set to basic waveforms or to more complex sounds. These sources are then combined together

Figure 7.1 The original vector synthesizer from Sequential Circuits—Prophet VS.

at various levels using a joystick or other modulator. Even though the mix process could have easily been developed as a square with four corners, it was designed as a diamond shape, which purportedly led to the Prophet VS nearly being called the Prophet Diamond.

Each stage of the VS is controlled by envelopes that have more detailed stages than the traditional ADSR and can help craft very interesting sounds. The modulation matrix is also very powerful and can be used to synthesize sounds that are exciting and full of complexity. Here are the modulation parameters available on the original Prophet VS:

Sources

LFO 1
LFO 2
Pressure
Velocity

Figure 7.2 Close-up on the Prophet Diamond.

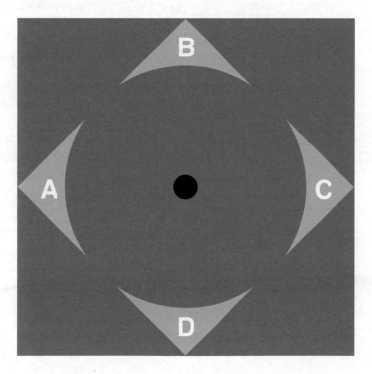

Keyboard
Filter Envelope
Mod Wheel

Destinations

Frequency
Filter
Mix
LFO 1
LFO 2
Amplitude
Pan
Chorus

An example of a modulation patch that demonstrates the power of VS is the mix destination, which lets the mix (joystick) be controlled between A–C and B–D. Attach LFO 1 to the mix between A and C, and then modulate the rate of LFO 1 by the modulation wheel. The sounds created in this situation are unlike anything else and are difficult to create using any other synthesis types. Shifting the mix between sounds using LFOs is an incredible way to make new sounds that can be as harsh as desired or rich and mellow.

The envelopes, which can be assigned to a variety of parameters, are also slightly nontraditional because they have four stages and each has a level and a rate. Many envelopes use a single level for the sustain and rates for the rest, but the VS has a rate and level for each stage, which offers the ultimate in flexibility.

The Prophet VS has never been an easy instrument to program because of its single data slider, but it is quite flexible if you have the patience. Even VS instruments that came after the original almost all use similarly difficult programming methods and haven't done any favors for musicians. VS synthesis is a very powerful tool for live performance, but patches typically have to be completely prepared in advance.

Once Sequential Circuits closed, Yamaha continued work on VS and eventually Korg also released several instruments.

Figure 7.3 Wavestation vector motion.

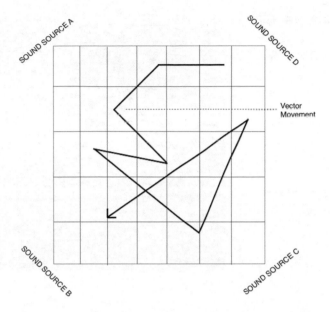

Each new iteration brought new features that enhanced the VS process and make each new generation unique and valuable. The Yamaha instruments introduced joystick tracking that could record mix automation and also added integration with FM synthesis. Korg introduced multisampled sources and wave sequencing that added sample playback for mixing full music loops in as one or more of the four sound sources.

VS has relied on digital technology since its beginnings, and, with the proliferation of software synths, VS has continued as a popular technique. Virtual instruments have given it new life and DAWs can be used with non-VS instruments to achieve the same general effect. Logic Pro X has two instruments with vector capabilities: the ES2 and Sculpture.

ES2

The ES2 has a semitraditional vector option with an X-Y pad that mixes various parameters, but it can also be programmed using a sequencer to create vector-controlled patterns. In addition to the vector module, the three oscillators are also mapped into a mix "triangle" used for combining the

247

Figure 7.4 Logic Pro X—ES2.

sources. Thus, the ES2 has two levels of VS combined with subtractive synthesis and FM synthesis to create one of the most powerful software synths available.

The ES2 is made even more powerful because it exists inside of Logic Pro X, with access to complex arpeggiators and other MIDI/audio effects. This system allows devices such as the Leap Motion to connect through Logic to the ES2 and uses hand motions to generate MIDI CC data. Instead of needing a joystick, it is possible to wave your hand in three dimensions to morph between the different parameters.

Sculpture

Another instrument in Logic Pro X is Sculpture, which is a much different type of instrument and relies on a modeled string as a sound source. It has a five-node, vector-like mod section, which gives the user the ability to mix between five different sounds. Instead of simply using five different sounds or wave shapes, each of the five nodes represents full instrument presets that include effects, filters, string settings, and so much more. Sculpture is just as much of a synth as it

Figure 7.5 Logic Pro X—Sculpture.

is a modeled instrument, and it demonstrates the power of vector synthesis in working with old and new musical technologies in creating very useable sound textures. The Leap Motion can also control the morph pad in Sculpture, which means it can not only create very interesting sounds, but can also be performed in a very expressive manner.

Figure 7.6 Sculpture's morph tool.

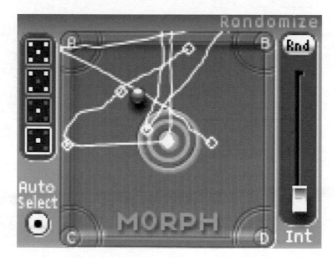

There is very little doubt that Sculpture's morph features are based on the concepts introduced with the Prophet VS, and, due to a complex interface with too much clutter, Sculpture is almost as difficult to program. Unfortunately, time hasn't made a difference in our ability to create easy–to-use instruments that are also complex, but if mastered, both the Prophet VS and Sculpture are capable of incredibly complex sounds that can be very expressive.

Creative DAW Control

It is possible with many digital audio workstations to create a vector synth-like experience by controlling faders from multiple sound sources. While this is perhaps an oversimplification of vector synthesis, there is very little difference from the original instruments. Pro Tools is one of the few DAWs that is still restricting custom control surfaces and it still isn't possible to use their surfaces in nontraditional ways, but systems like Logic Pro and Ableton Live are very flexible in the ways you can use MIDI controllers and OSC.

A default instrument in Pro Tools called Xpand 2 has four sound sources that can be loaded with different sampled instruments. Each sound has its own mix level and can easily

Figure 7.7 Sounds sources in Pro Tool's Xpand 2.

Figure 7.8 Track Stacks in Logic Pro X.

be assigned to incoming MIDI control data using Xpand's MIDI mapping feature. This is an easy way to create a vector synthesis-style environment. By using creative MIDI techniques, it is possible to create a map that even lets a joystick control the various levels.

An example of a powerful way to work with Logic Pro X and a VS type of experience uses Track Stacks, which let you combine multiple instruments into a single "track" that responds to a single MIDI input, and yet you can still modulate each instrument separately. In Logic, you can control volume levels of individual instruments using MIDI control data and so it is relatively easy to attach an external device such as an iPad or Leap Motion to mix the different instruments.

iPad Control

It's hard to beat a joystick in terms of performability when it comes to controlling the vector, but a multitouch interface does something special because it can create multiple X-Y controls that can be used at the same time. An app such as Touch OSC or Lemur can be programmed to create performance data for more than just volume/mix relationships, and you are only limited by the number of fingers you have and by your imagination.

If the app permits you to create a custom interface, then think about what you could do with 6 or 12 X-Y pads all attached to parameters on different instruments. You can play the instrument using a keyboard in one hand and then control an incredible amount of parameters using the other hand. In a recording studio setting, you can record the MIDI track with

Figure 7.9 iPad Running Lemur.

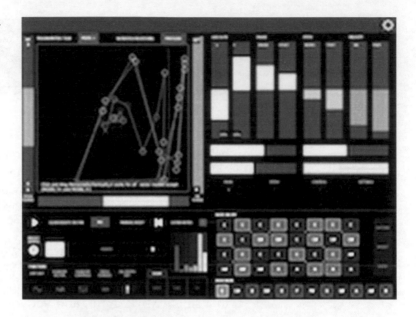

both hands and then use both hands to perform an additional layer of modulation.

Programming Vector Synths

This next section explores programming using many of the instruments discussed above. The primary objective is to give you an idea of how each actually works so that you can decide if one of these belongs in your instrument collection, or perhaps you'll feel like avoiding them completely. The best news is that a virtual replica of the Prophet VS is now available as an iOS iPad app, which means that the iPad collection is continuously growing and you should consider buying one. This app's OS X version is used below to demonstrate programming the original VS instrument.

Prophet V

The Prophet V instrument from Arturia improves on two Sequential Circuit originals in all of the right places and creates a massive, new instrument that uses both. The Prophet V

Figure 7.10 Arturia's Prophet V.

combines the first and last synths made by Sequential, the Prophet 5, and the Prophet VS. You can use the instrument in either mode, or as a combination of them to create a very powerful, new instrument. For this example, only the Prophet VS portion is used to demonstrate the programming techniques of the original VS instrument.

The first step is to choose the waveforms and tuning of the four sound sources. The first time you do this, you'll likely feel a little lost in knowing what to do—but after you go through the entire process, it will become more clear how to set the initial sounds. It is important to get used to sweeping through with the joystick, because if you don't, then you'll miss the core elements of the sound. Make small adjustments and then move the joystick to see how the new settings play with the other sounds.

For this example, four basic wave shapes are used, even though other more complex sounds are also available. Once the sounds are selected, then the tuning, envelopes, and filters are set. This portion of the process is very similar to creating in subtractive synthesis. After that, modulators are available to further change your performance experience. For this example, an LFO is used to modulate the mix parameter along just

Figure 7.11 Close-up of oscillators.

Figure 7.12 Digital waveforms.

one axis. The sources are tuned together but an octave apart so that the LFO rhythmically switches between them and creates a sequence-like pattern. The other axis doesn't have this same pattern, so it is possible to turn it on or off by moving the joystick to a different quadrant.

This sample patch demonstrates the ability of vector synthesis to take existing sound creation tools and make an instrument that is more expressive, flexible, and complex. It is possible to use other synths to create all of the same sounds, but there is

Figure 7.13 The Prophet V modulation matrix.

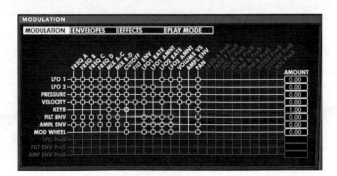

Figure 7.14 Prophet V LFOs.

something easy about the way vector tools work in combing certain sounds together that makes it unique and a certifiably different type of synthesis.

If you use the same LFO example above and apply it to each individual axis and create rhythms in both directions, but use different tempos for the LFOs, then you'll begin to see the complexity of using vectors in situations that result in unexpected sounds that are unpredictable. Each axis can be treated as completely different patches that are subsequently mixed between, just as a DJ might mix together two completely unrelated LPs. The key is to be creative and willing to explore new sounds as you search for the perfect mix of various elements.

Korg Wavestation

The original Wavestation was released in 1990 and picked up with vector synthesis where the Prophet VS had left off. At the time, it represented a bridge between traditional synthesis and an ever-expanding sonic experience. You have all heard the Wavestation in both pop music tracks and in an unlikely place when you hear an Apple computer turn on. Jim Reekes claims he used the Wavestation to create the startup sound

Figure 7.15 Korg Wavestation A/D.

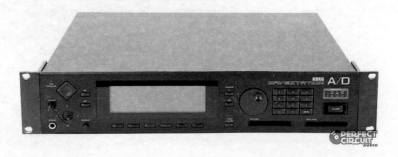

used first in the Quadra 700 (1991) and has been used ever since. If you own a Wavestation, then try the Sandman factory preset as it apparently is a slightly modified version of the sound, played as a stretched, C major chord.

The Wavestation SR came after several other Wavestation releases and is one of the most powerful vector synthesizers ever made, but also one of the most difficult to program. It is an example of a device that clearly wants to be a preset playback instrument while still maintaining all of the power of what the original Prophet VS and other in-between instruments could do. For this example, let's look at a performance patch that has a maximum part limit of eight different sounds. In addition to the four-source mix vector, the eight parts in the performance patch can be split across the keyboard or by velocity range. This means that you could have a patch that has bass in the lower keys, a piano in the middle keys, and a lead synth on the high keys, and when you play harder/at a higher velocity, then all of the sounds switch to something different.

The Korg SR uses PCM waves as the sound source for each patch. It has 484 individual internal waveforms and more can be loaded with expansion cards. Each voice has access to effects, envelopes, and modulation. The hardest part about programming the SR is that you have a tiny screen and very few buttons, and yet there are so many pages of editable data. There is even a vocoder, which is something Korg is known for on many of their instruments.

For this example, we are using very similar sounds to the example above with the Prophet VS and you can see how difficult it can be using the buttons on the front face. The first

Figure 7.16 Korg Wavestation SR LCD screen.

step is to initialize a patch so we can start from scratch. We can experiment by switching out the oscillators with various waveforms, but because there are 484, we have an incredibly large pile to search through. The SR can also place these together into a wave sequence and can create a whole different world of sounds and combinations.

The SR does not have a joystick, but you can attach a MIDI joystick or another MIDI source with adjustable controller data. By default, the SR is set to CC 16 and CC 17 for this control and so you can set your MIDI controller's faders to these or change the settings on the SR to whatever your equipment is set to. Once again, the Leap Motion or iOS apps can also easily be set to these and are able to control the vector mix. It may be best to use a DAW to manage the MIDI data because it can combine different inputs into a single output to control the SR. In Pro Tools, you can use a MIDI track with the input set to "All" and the output sending to the SR. The same is true for Logic Pro or nearly every other DAW.

Korg Wave Sequencing

Vector synthesis and several other types of synthesis have crossed paths in several specific ways, and so the lines are gray concerning what type of synthesis is actually taking place. Wave sequencing on the Wavestation uses a series of wave files that play in order as any of the four "oscillator" sources. This is very similar to wavetable synthesis, which also has access to a series of sounds that can be manipulated in a very similar fashion. Wave sequencing has not expanded outside of Korg instruments, but it has continued to be available in more recent Korg releases, such as their OASYS instrument and the amazing Korg Kronos. The one place where

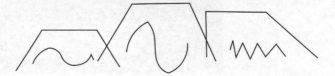

Figure 7.17 Illustration of a three-step wave sequence based on drawings in the original Wavestation manual.

wave sequencing hasn't appeared yet is in the wildly popular Korg Gadgets app for iOS.

Wave sequencing is a technique that Korg introduced with the Wavestation, but it is not integral to the vector synthesis engine. The performance manual describes the wave sequencing process in enough detail that it seems they are making sure no one confuses it with wavetable synthesis, but the similarities are remarkable. Here are the basic specifications of wave sequencing, which you can compare to wavetables in the wavetable synthesis chapter.

1. Memory banks have 32 wave sequences (0–31).
2. Each sequence has up to 255 individual steps.
3. Banks have a maximum of 500 steps.
4. Steps and step ranges can be looped.
5. The initial step and step motion can be modulated.
6. Layering patches in performance mode means that up to 32 wave sequences can be triggered simultaneously.
7. Wave sequences can be populated into each of the four sound sources, for full vector synthesis mixing.

Figure 7.18 Korg Gadget for iOS.

Prophet VS Patches in the Wavestation

The sound of the original Prophet VS is unique, partially because of the wave shapes included, but also because of the limitations and quirks of the early digital audio capabilities. If you want to explore the waveforms of the original Prophet VS, then all you need is a Wavestation, which has them in its memory as stored waves. These original waves are reproduced and used in patches and performances. They are labeled VSxxx, with the xxx being an associated number for each wave.

If for no other reason, it is worth buying a Wavestation or Wavestation SR just to have the experience of working with these original waves in a hardware-based system that is both difficult to program and yet full of rewarding results. The biggest difference is that the Wavestation instruments have no analog components and different filters/effects. This means that even if you program them as closely as possible, there will still be some significant differences in sound color.

Effects and Multisets

Another consideration to think about with the Wavestation is that it has the ability to perform using performance settings and multiset patches. The performance option lets individual patches to be mixed together to create an all-inclusive instrument. The multisets allow 16 patches to be accessed independently on each of the 16 standard MIDI channels. This sounds like it is a powerful way to work because it essentially gives you 16 Wavestations for programming, but the primary limitation is that there are still only two global effects that cannot be accessed individually by each of the 16 patches. Even in performance mode, you lose access to the effects for each layered sound, but performances are often designed with this in mind and so the effects work within this framework. Either way, it is recommended that you use the Wavestation as a single patch/performance tool, and avoid trying to get 16 patches working simultaneously.

ES2

The ES2 is both easier to program and more difficult at the same time. The advantage list includes a graphical interface and the ability to look at all of the parameters at once. The disadvantage list includes a difficult-to-understand interface with few labels and poor instructions. Instead of having four clear sources, the X-Y matrix is attached to various parameters and/or the oscillator level mixer. This means you can modulate filter frequencies and oscillator mix parameters simultaneously using timing nodes on the interface.

In this example, we are creating a simple pattern that affects the cut-off frequency and a mix between oscillator one and white noise. Instead of a joystick, we are using the matrix timing line to set when the X-Y parameters will change. We could use milliseconds or a subdivision of the project tempo grid, and so we are using 1/8 note variations.

When a node is selected, you can alter the settings for the cut-off and the oscillator mix amount before moving on to the next one and changing its settings. Turn off solo mode before testing the patch to see if you are happy. There are so many additional parameters available for tweaking, such as loop settings and X-Y mode settings. While there is no joystick and only three oscillators, this is clearly based on the original VS instruments.

Here's an example of connecting an iPad to control the X-Y parameters in real-time instead of using a defined sequence/tempo.

Figure 7.19 Vector view in Logic Pro's ES2.

Figure 7.20 Setting the grid resolution.

Figure 7.21 X-Y assignments.

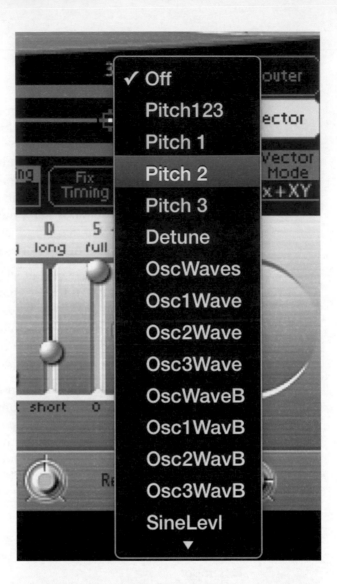

Figure 7.22 Smart control routing to X-Y.

Sculpture and Leap Motion

The primary part of this example is to demonstrate how to hook up the Leap Motion controller to modulate between the five nodes in Sculpture. You'll need the basic drivers installed, the GecoMIDI app for converting Leap data to MIDI (or equivalent software), and information about which MIDI data Logic Pro is expecting when it comes to controlling Sculpture. This last part always seems to be the hardest to figure out for each DAW, but all it takes is knowing where to look in Logic: this is highlighted in the accompanying image.

In GecoMIDI, it is easiest to use the "Up & Down Position" along with the "Left & Right Position" to control the modulation. Looking at the GecoMIDI interface, you'll see much more than that available, but in most cases, it is important to keep it as simple as possible because the technology is still

Figure 7.23 GecoMIDI.

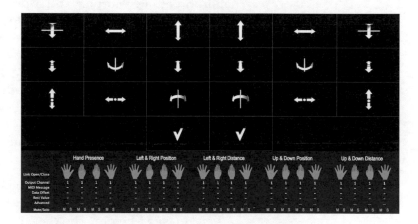

Figure 7.24 Logic's environment window with MIDI viewer.

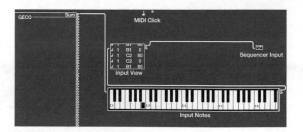

imperfect, and complex hand motions are not always interpreted correctly by the software. One of the easiest ways to decide which hand motions are the best for you is to wave your hand around and see what GecoMIDI sees the most clearly. Down below, you can set the MIDI CC data and that will be sent to your DAW by default when the GecoMIDI app is running. If Logic Pro isn't receiving the data, then check the preferences in GecoMIDI and troubleshoot online in the user forums. The GecoMIDI data will be recorded into your track if you engage recording mode on the initial pass, or you can record keyboard information first and then record a pass of Leap Motion second. There is an option in Logic to automatically combine MIDI regions when recorded on top of one another or if put in loop mode, so it won't be an issue to record everything separately.

In the case of Sculpture, the parameters that control the morph pad are MorphX and MorphY, which are controlled by CC 322 and CC 323. The MIDI specification doesn't technically extend beyond the default 128 controllers, and so calling them CCs is a stretch but this is how Logic deals with instruments having many more than 128 different parameters. GecoMIDI only works with the traditional 128 controllers, so you have to use the Smart Control system to map the output of GecoMIDI to the morph controls. The Smart Controls are a very easy way to route MIDI data from external sources to internal destinations.

Once the Leap Motion is configured, then there is a learning curve to performing because it unlike anything you've ever "played" before. There isn't any physical feedback, so

Figure 7.25 Sculpture's smart control assignments.

Figure 7.26 Leap Motion in action.

you have to learn to move your hand around with imaginary boundaries. The parameters in GecoMIDI can be adjusted for sensitivity and boundaries, but the default settings are satisfactory for most situations. Expect, however, to spend a number of hours getting used to it before using it live or in a pressure situation.

Vector Synthesis "Evolved"

Most synth types are becoming less independent in modern software implementations. A full-featured subtractive synth is likely to have FM, additive options, and even vector synthesis-style oscillator mixing. When Dave Smith began producing his own instruments again under the name Dave Smith Instruments, he released a hybrid digital/analog synth called the Evolver. This little beast has two analog oscillators and two digital oscillators . . . which should sound very familiar in the context of this chapter. Since its release, we have seen a number of instruments from Dave Smith, and most of them have embraced digital technology in ways many people did not expect.

The Evolver is a lesson on mixing sounds from different worlds into a single instrument that can speak both languages. This is what vector synthesis is all about and Dave Smith Instruments continues that tradition by continuously expanding the definition of what a synthesizer is in a modern context. It honestly doesn't seem like he is interested in calling a synth digital or analog, but by using the best of both worlds, he can create an instrument that is capable of more than either by themselves.

Korg Gadget iOS App

Once again, Korg is pushing the boundaries of music creation by developing and releasing a series of iOS music apps. At first they were most interested in creating mobile versions of classic hardware instruments, but, more recently, they have started to move into releasing new instruments that have been designed with new ways of doing things. Korg Gadget is a perfect example of this with a newly developed sequencer that follows in the tradition of Ableton Live and includes a large number of sound-source modules.

Among these modules is a newly designed vector synth called Kiev. Even though it is called a spatial digital synthesizer, it is in fact a vector synthesizer and it demonstrates the key elements in vector synthesis in a very streamlined design. For this next section, Kiev is explored as perhaps the most recently designed vector synthesis available and released in a multitouch experience.

Figure 7.27 Kiev module in Korg Gadget.

Kiev Interface

The bright yellow face is designed to look like an old piece of submarine sonar equipment or something similar. Almost immediately noticeable are the orange lights that move back and forth/up and down along the OSC MORPHER in the center of the device. In the top left corner is a menu to select between various presets, called Sound Programs. There are 41 provided and a place to save user presets.

There are four oscillators, each with basic pitch and tuning settings. Each oscillator can be assigned to one of 92 waveforms, some of which resemble traditional, simple waveforms, while others are substantially more complex.

Kiev Modulation

The Morpher is exactly what you would expect in a vector synth and it mixes between the four oscillators, labeled A–D. There are modulators hardwired to A/B and C/D, which affect the mix between them using an LFO model. You can adjust the depth and speed of the modulation, but not the waveform or other parameters. The lights on the edge of the Morpher represent the modulation speed between the oscillators.

Figure 7.28 Morpher Interface.

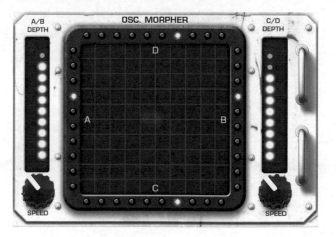

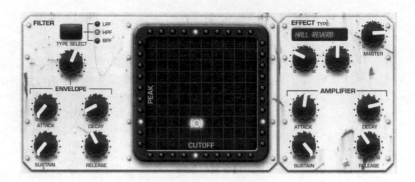

Figure 7.29 Filter and amp settings.

Kiev Filter/Amp

Press the button at the top of the instrument and it switches into the Filter/Amp mode. The filter can be switched between low pass, high pass, and band pass modes. There is a four-stage envelope, which is activated by adjusting the depth knob and is fairly straightforward. The Morpher pad switches from controlling the oscillator mix to controlling the peak and cut-off frequencies. This is a very intuitive way to adjust the filter settings and can even be used during a live performance.

The effects section provides access to a single effect engine, which can be assigned to any one of 25 different effects. These range from reverbs to delays, compressors, distortion units, and equalizers. Each effect is customizable using two knobs, which unfortunately are not labeled as to what they do. This results in a little trial and error, but that is also something that can be fun.

The interface also includes a master volume and an amplitude envelope. The keyboard below is barely playable, but at least you can assign a custom scale to make up for the small size and you can change the available octave.

Kiev MIDI Mapping

One of the best parts about Korg Gadget is that you can create a custom MIDI map for controlling all parameters. This

Figure 7.30 Gadget MIDI
mapping options.

means you can connect an external controller and easily manipulate all of the instrument functions. There are multiple ways to connect a controller to Gadget and not all of them are created equal.

1. Camera Kit—The camera connection kit is a pretty reliable way to connect a MIDI controller to the iPad as it uses a USB cable and if the controller is class compliant, then it should automatically be recognized and work. Class compliancy is a standard that has been agreed to by developers that lets hardware and software work together without any additional drivers being required. Essentially, the needed drivers

Figure 7.31 Apple iPad camera kit.

Figure 7.32 OSX audio MIDI setup utility.

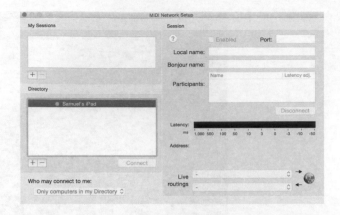

Figure 7.33 Bluetooth MIDI.

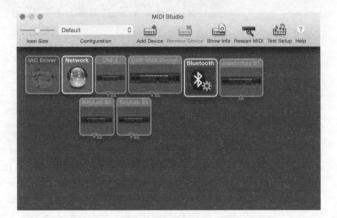

and software are already made available. The primary limitations include low power availability and it ties up the only port on the iPad, which would no longer be available for charging the iPad. A solution for powering a MIDI keyboard is to connect a powered USB hub to the camera kit and then the keyboard to the hub.

2. Wi-Fi Connection—If you are on the same wireless network, then you can connect using MIDI over the network, but the reliability of this system is based on network traffic and is prone to poor latency response times.

3. Bluetooth MIDI—iOS 8 and OS X Yosemite include a new option to communicate with MIDI over Bluetooth. The

initial implementation has not been trouble free and it often takes a fair amount of time to set it up in a working state. Once it is configured and working, it is very solid and very responsive. Hopefully, Apple continues to develop this and further enhance its reliability.

Using a DAW such as Logic Pro X in conjunction with Gadget, it is once again possible to attach other external controllers such as the Leap Motion for X-Y control. You might wonder why someone would want to use Gadget if they already have Logic Pro X and that might be a legitimate point, but there is something very captivating about having a full production studio in a tablet thinner than a pencil that you can take with you everywhere you go. It is also nice that you can still plug it into a bigger production studio and use the larger controllers and MIDI tools available in a full-featured DAW. The only missing piece is getting full quality audio in/out of the iPad and into your DAW. It is certainly going to arrive at some point in the near future.

Kiev Summary

This is an incredibly limited vector synthesizer, but it shows a refreshing trend away from the past 15 years, where every new instrument had to have everything possible under the sun. Korg, with all of their experience in vector synthesis, decided that four sound sources, two hardwired modulation LFOs, a filter, a single effects bank, and an output envelope are everything needed at the core of this instrument. Even wave sequencing has been omitted from this latest iteration—not because of the iOS format, but because it is clear from some of the other modules in Gadget that it could have been accomplished. How does Kiev sound? It is capable of creating very complex sounds that are exciting because they are never stagnate and remain very intuitive for live performance. The main thing I wish Kiev had is the ability to import custom waveforms for creating your own patches.

Vector Synthesis Summary

Of all of the types of synthesis covered in this book, it often feels like vector synthesis is the least innovative or the one that is least able to stand on its own. Even at its creation, it was as much of an experiment with digital synthesis as it was an exploration of vectors. It is extremely easy to accomplish the same end results of mixing sources together using a number of other techniques.

At the time it was packaged, it made a big enough impact that it has remained a contributing synthesis tool all of these years. Very few companies are currently releasing new instruments with an exact replica of vector synthesis as it was originally designed, and I wonder if it is because the rights to doing so are still covered by patents. Arturia has released Prophet VS clones, and Korg continues to work with vectors, but exciting new releases from MOTU (MachFive) and others have left it off as a primary synthesis type.

The current state of instruments leads us to believe that vector synthesis is less a type of actual synthesis and more of a vehicle for synth engines to accomplish certain modulation tasks. You won't see a diamond grid on every instrument, but you will often have the ability to mix between two patches or automate mix levels for various sounds. The original Prophet VS taught us a different way to think about programming sounds that has evolved over time and morph into new timbres.

COMBINATION SYNTHESIS AND MODELING 8

Up to now, all of the synthesis formats discussed in this book have been stand-alone, concrete formats that follow a predictable path and feature set rules and guidelines pertaining to how they are designed. This chapter introduces synthesis formats that do not necessarily follow such a rigid structure. The landscape that physical modeling, emulation, and combination synthesizers exist in is murky at best. In fact, the term *combination synthesis* is used more as a catch-all phrase for synthesizers that do not fit an exact mold and are difficult to categorize. That being said, there are some rigid ideas behind physical modeling synthesis, but due to its prevalence in workstation synthesizers and virtual instruments, the strict forms that are used to categorize synthesizers are oftentimes thrown out the window when dealing with physical modeling instruments. Hopefully this chapter will not only aid in clearing up a few misconceptions about the powerful synthesizers that exist in this grey area of synthesis, but also show them in a new light by detailing their strengths.

Introduction to Physical Modeling

Physical modeling synthesis, or PM for short, is the act of re-creating sound through algorithms which emulate physical properties. For instance, a physical modeling synthesizer that is modeling a stringed instrument, such as a cello, will have algorithms and formulas in place to mimic the way the bow is pulled across the strings as well as the way in which the body of the instrument vibrates while being played. Physical modeling synthesis will go as far as having algorithms in place that allow users to adjust the density of the wood of a cello, as well as how the bridge sits on the surface of it. PM synthesis can be thought of as a behaviorally based synthesis technique, in that it mimics the behavior of various components of physical instruments. In essence, the behavior of the wood, strings, bridge, pegs, body, and way it's played are all re-created through complex mathematical formulas and equations. Physical modeling synthesis is an extremely powerful synthesis tool that allows users to re-create the sounds of physical instruments with much more control and dexterity than might be allowed with sample-based synthesizers. Physical modeling synthesis is most often associated with orchestral stringed instruments, but it is technically capable of emulating any type of instrument or sound. Many PM synthesizers will model percussion instruments or mallet instruments such as marimbas and xylophones as well.

Although both physical modeling synthesis and sample-based synthesis attempt to re-create an instrument or sound, there is actually quite a large difference between the two technologies. Sample-based synthesis uses any number of recordings of the instrument it is trying to mimic as a starting point in its tone creation process. By using recordings, the user is basically stuck with the individual instrument and environment used when creating the recordings. Obviously, the user has some freedom such as reverberation and pitch shifting in order to change the sound, but for the most part, the user is stuck with the instrument recorded. Physical modeling synthesis, on the other hand, frees the user from being stuck with a single instrument or environment. This freedom comes because PM synthesis does not use recordings as the base of its sound-creation process. Instead, formulas and algorithms are created based off of the physical attributes of the materials that make up the instrument and environment. Although this book will not make an attempt to go into the various laws of physics taken into consideration by PM synthesis designers and developers, it is safe to say that the way in which wood, metal, or plastic interacts with its environment when struck, plucked, bowed, or strummed is painstakingly analyzed and then coded into the PM synthesizer's software.

Many people might rightfully assume that a physical modeling synthesizer is unnecessary in today's ever-expanding music technology environment. Many could argue that since it is possible to buy huge sound libraries that contain hundreds if not thousands of samples of just one type of instrument, physical modeling is not necessary. This statement however, is narrow-minded and only focuses on physical modeling synthesis's ability to *faithfully* re-create instruments and pays no heed to arguably the most powerful aspect of PM synthesis: its ability to *unfaithfully* re-create instruments. When using PM synthesis, not only is the user free to change the makeup of the instrument in order to change from a dark wooded instrument to a light wooded instrument, but the user is free to change properties that might not be possible, or advisable, in the physical world. For example, some PM synthesizers might allow for creating cellos made of metal and played by felt mallets being struck against its strings. The possibilities are endless in reference to the crazy and inspiring sounds that can be thought up and created using physical modeling synthesis.

Key Features of Physical Modeling Synthesis

Because physical modeling synthesis does not adhere to the strict rules and regulations found in various other synthesis formats, each individual PM synthesizer is quite unique from any other PM synthesizer. We will explore some of the most famous and current physical modeling synthesizers in depth later in the chapter, but it is important to first focus on some of the controls one might expect to find on a physical modeling synthesizer. Because the controls and adjustable parameters will change based off of the instrument it is designed to model, we will divide this exploration into three categories—stringed instruments, nontuned percussion instruments, and, finally, tuned percussion instruments.

Stringed Instruments

Perhaps the most prevalent category of instrument modeled for PM synthesis is stringed instruments. Stringed instruments include the range of stringed orchestral instruments such as cellos, violas, violins, and basses as well guitars, bass guitars, and even harpsichords and world instruments such as Zithers.

String Material

Although each physical modeling string synthesizer will have its own unique set of controls, a few controls are more common than others. Firstly, a way of controlling the strings themselves will typically be available. This type of control includes the ability to change strings from steel to nylon, or even change how hard they are pressed by the finger or damping device. Finally, the tuning of each string, as well as the harmonic overtones it produces will typically be adjustable by the user in order to create an instrument that is perfectly in tune, from its highest harmonic to a more realistic instrument that has some out-of-tune harmonics, all the way up to a completely out-of-whack experimental instrument.

String Excitation

The next most common type of control found is known as *exciter* controls. An exciter control relates to the way in which the instrument is played or how its strings are set into vibration. Typically, a PM string synthesizer will allow users to choose from a number of ways in which to excite the string such as bowing, finger plucking, picking, or smacking with a mallet or hammer. The user will not only have control over how hard the string is excited, but will typically be able to

Figure 8.1 Various methods of exciting a string.

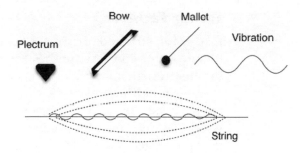

designate where the string is struck, such as above or below the string or higher or lower on the body.

Physical Makeup

In addition to designating the material that strings are made of and the way in which they are excited, users will also be able to designate the material and physical properties that make up the body of the instrument. The physical makeup section of a PM synthesizer will usually have a huge amount of parameters that the user can adjust. Some examples of these parameters are the overall size of the instrument, the material it is made up of, how thick the instrument as a whole is, and, finally, the thickness of the material that makes up the instrument. Changing the physical attributes of the instrument's body will drastically change the overall timbre of the sound.

Tuned Percussion Instruments

The next type of PM synthesizers we will examine model tuned percussion instruments. These instruments range from pianos to mallets and xylophones, all the way to tubular bells and beyond. As with the modeling of stringed instruments, PM synthesizers that model tuned percussion instruments will vary greatly between instruments. That being said, we will examine a few of the controls one might expect to run into.

Exciter Section

Like with stringed instruments, the exciter section of a tuned percussion PM synthesizer will designate what sets the instrument into vibration. Users will usually be able to choose between exciters such as beaters, mallets, hammers, and sticks as well as the material they are made up of. Users can designate the exciter with extreme precision and mix and match materials to make new types of exciters that might not be common in the physical world, such as metal mallets with felt tips. Additionally, it is not uncommon for users to designate not only the exciter's material, but also how stiff it is.

Resonator Section

The next most common thing users will be able to determine on a PM-tuned percussion synthesizer is its resonator—or what is being set into vibration on the instrument. Choices for the user include strings, plates, beams, membranes, marimba beams, pipes, and tubes, as well as a variety of other, less common resonator sources. Like when modeling stringed

Figure 8.2 Example of a resonator.

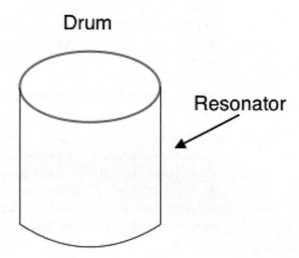

instruments, changing the resonator material will result in drastic changes to the timbre of the sound.

Nontuned Percussion Instruments

Tuned and nontuned percussion instruments are quite similar and oftentimes have a lot of overlap, but they merit a separate discussion. Nontuned percussion instruments involve drums, save for tom toms, cymbals, bongos, congas, and various other percussion instruments. Although individual drums can be "tuned," they are typically not tuned to individual pitches and therefore are not considered tuned percussion instruments. Like with tuned percussion and stringed instruments, users will often be able to designate the material and stiffness of the exciter as well as the material and size of the resonator. Many PM percussion synthesizers will allow users to choose the material of the drum cavity as well as its inherent thickness.

Like with granular synthesis, many separate synthesis forms are lumped together under the phrase of *physical modeling*. These separate but related synthesis formats are waveguide synthesis; modal synthesis; McIntyre, Schumacher, and Woodhouse synthesis; Karplus-Strong synthesis; and formant synthesis. These other synthesis formats can be thought of as different approaches for physical modeling synthesis. Although they are all closely related, it is important to have a brief understanding of each format.

Figure 8.3 Digital waveguide synthesis attempts to mimic traveling sound waves through various mediums.

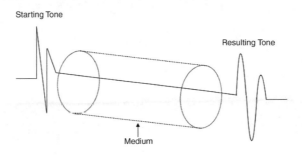

Starting Tone

Resulting Tone

Medium

Waveguide Synthesis

Digital waveguide synthesis is a form of synthesis that attempts to mimic the way in which sound waves travel through a physical medium. For example, think of a clarinet. Once vibrations are created at the reed of a clarinet, sound waves propagate through the tube of the instrument. The material in which the clarinet is made of as well as which tone holes are open or closed, all play a part in the overall timbre of the instrument. Digital waveguide synthesis utilizes complex mathematical formulas in order to replicate these functions. In essence, a waveguide can be thought of as the medium that guides the wave, such as the tube and bell of a clarinet. Digital waveguide synthesis is used heavily in most physical modeling synthesizers in order to accurately replicate physical mediums.

Figure 8.4 Modal synthesis focuses on the frequency domain of vibrating objects.

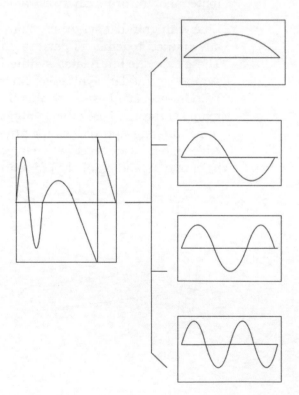

Modal Synthesis

Modal synthesis is one of the more complex techniques employed in physical modeling synthesis. Modal synthesis focuses on the frequency domain of a vibrating object. Modal synthesis can be crudely compared to additive or Fourier synthesis in that a vibrating object is deconstructed to base frequencies or modes that can then be analyzed and resynthesized. The formulas used in modal synthesis are complex and exist outside the scope of this book, but a basic understanding is helpful. In essence, modal synthesis is used to designate frequency information resultant of sound moving through various mediums. Returning back to the clarinet example, modal synthesis can be utilized to create a much more accurate representation of how pitch changes when tone holes are covered or opened. For this reason, modal synthesis is used heavily in physical modeling synthesizers.

McIntyre, Schumacher, and Woodhouse Synthesis

McIntyre, Schumacher, and Woodhouse synthesis, or MSW for short, is yet another technique employed in the

Figure 8.5 MSW synthesis focuses on the time domain of vibrating objects.

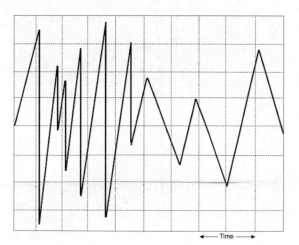

modern physical modeling synthesizer. While modal synthesis focuses on the frequency domain of vibrating objects, MSW synthesis focusses on the time domain of vibrating objects. Again returning back to the clarinet, MSW synthesis focuses on the linear and nonlinear excitation resulting from the user blowing on the clarinet's reed. An inherent pitch is present on any clarinet simply from the relation of the length of the instrument and the amount of blowing force applied to the reed. MSW synthesis is used to deconstruct this inherent pitch and then apply it to a resynthesized model.

Karplus-Strong Synthesis

Karplus-Strong synthesis is used in plucked string and drum PM synthesis. Karplus-Strong (KS) synthesis utilizes

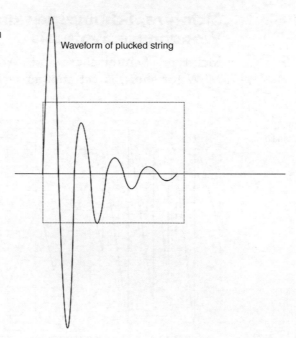

Figure 8.6 KS synthesis focuses on frequency and amplitude loss in plucked string over time.

Waveform of plucked string

waveguide synthesis for its algorithms. KS synthesis uses complex waveguides at the onset of every note and then sharply diminishes the frequency content in quick succession. By starting with a complex set of frequencies and then quickly diminishing that amount, plucked string and drum sounds can be created quite accurately. When a violinist plucks a string, for example, the sound is extremely rich at the onset and then quickly falls to a much simpler, sine wave-like tone once the higher order harmonics fade away. KS synthesis attempts to mimic this through complex algorithms and formulas.

Formant Synthesis

Formant synthesis involves passing signal through a number of complex resonant filters. In practice, formant synthesis is most often used in speech synthesis and creating vowel-type sounds. In its traditional sense, formant synthesis is not typically used outside of laboratories and universities. That being said, aspects of formant synthesis, such as formant filters, are used heavily in commercial physical modeling synthesis. Unlike the previous synthesis formats and algorithms examined, formant filters are used as a feature on PM synthesizers rather than a sound-generation source. Formant filters are often sought after for the unique and interesting sonic characteristics they impart on the overall sound.

Figure 8.7 Formant synthesis utilizes a series of filters known as formant filters to mimic individual formants found in acoustic instruments.

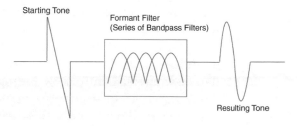

Starting Tone

Formant Filter
(Series of Bandpass Filters)

Resulting Tone

The Physical Modeling Model

These various synthesis techniques are all utilized to some extent in most modern physical modeling synthesizers. Because each technique is extremely well suited for its exact purpose, they all can be utilized to create an extremely powerful physical modeling synthesizer. Much like LEGO bricks, aspects of each of these formats can be used together in order to create a single instrument. Most PM synthesizers will not advertise that they use these types of algorithms nor will they offer the user the ability to turn on or off individual algorithms. Instead, all or most of these algorithms will be working "under the hood" of a PM synthesizer in order to create the complex and awesome sounds one desires.

Putting It All Together

Up to now, we have discussed the various attributes one can expect to find with a physical modeling synthesizer as well as the theory behind its numerous algorithms. In past chapters, once the discussion of individual parts is given, an explanation and examples are provided on how to build a sound from scratch. This is not the case with physical modeling synthesis. In reality, the algorithms used to achieve the breathtaking sounds physical modeling synthesis offers are just too complex to be useful to the average musician. Because of this, most physical modeling synthesizers will rely on factory presets as starting points and give users the ability to adjust certain parameters. Although this might seem prohibitive to some, it is really the only way in which physical modeling synthesis can be useful to the masses. So instead of starting new sounds completely from scratch, users will build up their sounds from the factory presets or factory default settings. Each PM synthesizer will allow for unique sound modification, but most users can expect to see a few standard parameter controls: including the ability to designate the material an instrument is made of, how the instrument is excited or played, and individual tensions and frictions of various parameters. Most PM synthesizers will also provide some standard synthesis adjustability such as pitch bending, modulation, and some type of filtering.

Every synthesis format discussed up until this point requires practice and patience in order to master it. Physical modeling synthesis is unique though in that it requires an extreme amount of practice to master. While most synthesizers can simply be plugged in and fiddled with to create useable sounds, PM synthesis requires extreme precision in order to create desirable sounds that are useful and not cheesy. Because PM synthesis emulates real instruments through algorithms representing physical attributes, extreme care must be taken when playing to ensure things like breath amount are interpreted as real sounding, rather than odd, digital artifacts. Sample-based synthesis is free from this challenge as things like breath noise is recorded into the samples. Unlike sample-based synthesis, PM synthesis requires the user to delicately add in things like breath noise when appropriate.

The amount of time and practice it takes to create realistic sounds using PM synthesis should not deter you. On the contrary, the time spent with a physical modeling instrument should be viewed as fun and engaging. To this day, despite all the advancements made in synthesis and sampling technology, physical modeling synthesis is still the most powerful way to emulate physical instruments. PM technology is still evolving and becoming better every time CPU power increases.

Hardware Physical Modeling Synthesizers

Now that we have discussed physical modeling synthesis and the technology behind it, let us examine some physical modeling synthesizers. By examining the various options users have when it comes to picking a PM synthesizer, we can explore some of the features unique to physical modeling synthesis.

Yamaha "VL" Line

Often cited as one of the earliest physical modeling synthesizers to hit the market, the VL-1 brought life back into the often stagnant and stock synthesizers of the day. With the VL-1, Yamaha introduced their own version of physical modeling technology, which they called Virtual Acoustic Technology. Yamaha shipped the VL-1s with a MIDI foot and breath controller to provide the most amount of control possible. With certain brass and woodwind patches, the breath controller would add more and more breath noise into the sound the harder one would blow. At extreme levels, pitch and timbre instability would arise that would oftentimes sound extremely musical and interesting. The VL-1 revolutionized the way in which musicians interacted with a synthesizer. For many users, playing the VL-1 made them feel that they were interacting with a real, physical instrument, rather than a keyboard-controlled computer: something that had not been felt since the days of analog subtractive synthesizers.

Due to the extreme complexity of the VL-1's sound-generation engine, Yamaha did not provide the possibility for users to create their own patches from scratch. Instead, Yamaha included a wealth of customization capabilities that could be applied to the factory set sounds. The user adjustment options included:

- Pressure (bow speed)
- Scream (a type of controllable chaotic distortion)
- Tonguing (emulates tongue dampening of reeds)
- Amplitude
- Breath Noise
- Throat Formant (emulates effects of lungs, mouth, and throat)
- Damping (emulates friction)
- Growl (LFO effected pressure control)
- Dynamic Filter (filter cutoff adjustment)
- Harmonic Enhancer
- Embouchure (emulates the tightness of a players lips)
- Vibrato

- Absorption (emulates natural high frequency loss)
- Pitch (affects overall pitch of the instrument)

All of these parameters could be mapped to any MIDI controller, allowing for use with traditional sliders and knobs as well as foot and breath controllers. So although Yamaha did not allow users to create their own sounds completely from scratch, the sound could be dramatically changed from the factory presets. Despite the extreme attraction many people had towards the VL-1, it was extremely cost prohibitive to many people. Yamaha then released the VL-7, which was more affordable. Although slightly less powerful than its older sibling the VL-1, the VL-7 was widely adopted by people excited about physical modeling technology.

Technics SX-WSA1

The Technics SX-WSA1 synthesizer is an elaboration on physical modeling synthesis. The Technics SX-WSA1 uses what it calls *acoustic modeling* rather than physical modeling. Both acoustic and physical modeling are extremely similar and do not necessarily warrant separate names. The engineers at Technics decided that they wanted to create a more accessible physical modeling synthesizer. They felt that synthesizers like the Yamaha VL-1 were too complex and unintuitive to be useful. When creating the SX-WSA1, the physical modeling model was examined in depth and deconstructed in order to make it simpler. The engineers at Technics decided that in order to accurately emulate physical sounds, they could split sound into three parts: drivers, resonators, and modifiers.

According to Technics, drivers are what produce raw sound like a piano hammer hitting a string, or the reed of a clarinet vibrating. Moving on, resonators are what color the sound emitting from the driver. An example of a resonator is the body of a musical instrument. Finally, the modifier is what shapes and colors sounds—mainly a filter. By breaking sounds down into these three categories, the engineers at Technics believed they could create a synthesizer that used drivers, resonators, and modifiers to create emulations of physical instruments in a much more user-friendly manner than the physical modeling-monster synths that came before it.

The SX-WSA1 provided users with a large number of drivers (over three hundred), and tons of resonators. Users could then combine various drivers with various resonators in order to build sounds. For the first time, users were free to create sounds from the ground up rather than starting at factory presets. Although starting with premade drivers and resonators is not really starting completely from scratch, it was leaps and bounds more customizable than the Yamaha VL platform. The SX-WSA1 proved to be a quite successful synthesizer and the new approach it offered to physical modeling synthesis paved the way to the physical modeling soft synths we know and love today.

Software Physical Modeling Synthesizers

Due to the complex nature of physical modeling synthesis algorithms, as well as the number of computations that must be made instantaneously, a physical modeling synthesizer is really only as good as its computational power. Therefore, physical modeling synthesis really excels in a software or VST environment where it can harness the user's computer for CPU power. There are a large number of software PM synthesizers on the market, so we will only discuss a few of them that offer unique and interesting features.

Ableton Live—Tension, Electric, Collision

Three physical modeling synthesizers are officially offered to Ableton Live users—Tension, Electric, and Collision. Tension acts as a physical modeling string synthesizer while Electric adheres to electric pianos and Collision focuses on drums and percussion. All three of these PM synthesizers are extremely powerful and are extremely impressive, especially when compared to their low price point.

Tension allows users to designate almost every attribute of a stringed instrument through a variety of PM algorithms. Using a similar *Acoustic Modeling* architecture to the Technics SX-WSA1, users adjust parameters of drivers, resonators, and modifiers. Starting with the drivers, Tension allows strings to be excited through the means of picking, bowing, plucking, or even being struck with a mallet or hammer. Next, users can choose the size, shape, thickness, and material of the body of the instrument. Combining various excitation and resonator models allows for anything from extremely accurate instrument emulation all the way to extremely bizarre and alien-type sounds. In addition to choosing drivers and resonators, users can also determine the amount of finger pressure applied to strings, as well as the amount of mechanical dampening from things like the felt on piano hammers.

Ableton's Electric is a physical modeling synthesizer designed to emulate electric pianos such as the famed Rhodes and Wurlitzer electric pianos of the 1970s. Electric follows a similar model to Tension in that it is based off the *Acoustic Modeling* model found in the Technics SX-WSA1. Electric allows for a wealth of user-adjustable control that relates directly to electric pianos. Some parameters that are able to be adjusted are:

- Mallets (how the tines are struck)
- Tines (tone producing mechanism in electric pianos)

Figure 8.8 Ableton Live's Tension PM string synthesizer.

Figure 8.9 Ableton Live's Electric PM keyboard synthesizer.

- Pickups (transducers that turn physical vibrations into electricity)
- Dampers (mechanism which stops the vibration of tines)

Electric is capable of extremely accurate electric piano emulation and, in most instances, behaves more accurately than even some of the best sample libraries on the market. Electric even gives the user the ability to change the sound of the electric piano to emulate the way age affects the instrument.

Ableton's final official physical modeling synthesizer is known as Collision. Collision is a drum and percussion emulation synthesizer that uses the same *Acoustic Modeling* model as Tension and Electric. Collision excels at mallet instruments such as marimbas, xylophones, and glockenspiels. Besides mallet instruments, Collision is capable of most drum and percussion emulation as well as some percussion stringed instruments like pianos and dulcimers. When programming on Collision, the user first selects the resonator, which in this case is both the physical playing surface as well as the resonating body. The choices of resonators are:

- Beam (emulates beams of different materials and sizes such as are found in xylophones)
- Marimba (form of beam that produces characteristic overtones of marimbas)
- String (emulates strings of varying size and material found in piano and dulcimers)
- Membrane (drum head with adjustments for material and size)
- Plate (acts as rectangular plate with adjustable material and size)
- Pipe (emulates a long cylinder with one open end and adjustable other end)
- Tube (emulates a long cylinder open at both ends)

Figure 8.10 Ableton Live's Collision PM drum synthesizer.

Together, Tension, Electric, and Collision offer some of the best physical modeling synthesis options available on the market. The user control, playability, low price point, and overall sound quality are second to none. These three physical modeling synthesizers truly demonstrate the power PM synthesis offers.

Sculpture

Logic Pro's Sculpture synthesizer is, in essence, an extremely powerful physical modeling and combination synthesizer (more on combination synthesis further in the chapter). When creating a sound in Sculpture, the first thing that must be determined is the material being used. Logic has introduced an exciting way of choosing materials in the form of a material square. The material square is a square in which each corner represents a different material—nylon, wood, steel, and glass. A ball is present in the square, which determines the material being heard, and can be moved at any point in the square allowing for any combination of the four materials imaginable. Besides just choosing the material of the instrument, Sculpture provides a huge number of exciters that range from traditional plucking and mallet strikes to symbiotic excitation.

Figure 8.11 Logic Pro's Sculpture PM synthesizer.

Although Ableton Live's Tension, Electric, and Collision PM synthesizers may be best suited for instrument emulation, Sculpture is best suited for creating new sounds via physical modeling synthesis. In fact, Sculpture is one of the first and only PM synthesizers designed to emulate not instruments themselves, but the way in which they interact with the environment. By focusing on individual aspects of physical instruments rather than the instrument as a whole, the user is free to create sounds completely unrestricted by the bounds of factory presets. Besides the extremely powerful physical modeling engine found on Sculpture, a few other features are present that make Sculpture even more powerful. To start with, Sculpture allows for a type of vector-like control over various aspects of the synthesizer. This vector-like control is known as Sculpture's morph pad and it contains five points in which morph-able parameters can be routed to. Once parameters are routed to individual morph points, the user can manually drag the placement ball across the points and seamlessly move from one parameter setting to another. The placement ball can also be fully automated as well. Finally, Sculpture features some traditional synthesis capabilities such as filtering, LFOs, and envelope generators. Sculpture is an extremely powerful physical modeling synthesizer and is one of the most intuitive and creative software synthesizers available on the market today.

Analog Modeling

Another, perhaps more popular, version of physical modeling synthesis is available in the form of analog modeling synthesis. Analog modeling, sometimes referred to as virtual analog, is a derivative of physical modeling synthesis. As the analog resurgence began to sound in the 1990s and 2000s, many synthesizer companies felt the pressure to start producing analog synthesizers. Because of the uncertainty of how long the analog resurgence might last, as well as the difficulty of convincing shareholders to rehash old technologies, many synthesizer companies began offering analog modeling and virtual analog synthesizers as a compromise.

An analog modeling synthesizer will use many of the same techniques found in traditional physical modeling synthesizers. However, the goal of analog modeling synthesis is not just to re-create the sound of a particular analog synthesizer, but to mimic the way in which the individual analog components act. By mimicking the analog components rather than the overall timbre, analog modeling synthesizers are capable of creating much more realistic and believable sounds when compared to a digital subtractive synthesizer.

For example, one complaint many users have with digital subtractive synthesizers is the stepping heard when adjusting a parameter such as a filter's cutoff frequency. This digital stepping is resultant of digital increments being cycled when performing a parameter sweep. In reference to the synthesizer's filter, the filter can be heard opening or closing at small increments rather than a smooth transition as would be heard in an analog synthesizer.

By modeling the behaviors of the various analog components, analog modeling synthesizers can come much closer in sound to their analog counterparts. The algorithms used in analog modeling synthesis are designed to mimic circuitry behavior rather than physical property behavior like in traditional physical modeling synthesis.

Many hardware synthesizers, as well as VST plug-in synthesizers, have been released that use this analog modeling technology. Perhaps the most famous synthesizer to use analog modeling technology is the Clavia Nord line, such as the Nord Lead and Nord Modular synthesizers.

Figure 8.12 Filter responses on a digital subtractive synthesizer (right) and analog modeling synthesizer (left); notice the stepping present in the digital subtractive synth.

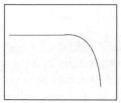 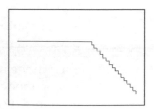

Figure 8.13 Clavia Nord Lead analog modeling synthesizer.

Released in 1995, the Clavia Nord Lead was a revolutionary new synthesizer that got a lot of people excited for the future of synthesis. The Nord Lead came at a time when subtractive synthesis was often overlooked and found only as afterthoughts on some combination digital synthesizers. For the first time in a long time, a hardware subtractive synthesizer was available that not only sounded great, but featured designated knobs for each parameter found on the synthesizer. The Nord Lead was capable of creating analog subtractive synth sounds, with full adjustability, without the hindrances of analog circuitry. The Nord Lead's great sound was created solely through analog modeling technology.

Reminiscent of an analog subtractive synthesizer, the Clavia Nord Lead features two discreet oscillators, a multifilter, two envelope generators, a fully featured LFO, and an amplifier: in essence, full subtractive bliss. Many diehard analog users either covet or at least respect the Clavia Nord Lead thanks to the brilliant sounds that can be achieved all through the means of analog modeling technology.

Like the Clavia Nord Lead synthesizer, another influential synthesizer, the Access Virus, utilizes analog modeling as a sound generation engine. First released in 1997, the Access Virus line of synthesizers has gone on to enjoy wide acclaim. Both the Clavia Nord Lead synths as well as the Access Virus line of synths can be heard extensively on Nine Inch Nails tracks and newer Depeche Mode albums to name a few.

Even in the height of the analog synthesizer resurgence, new analog modeling synthesizers are being offered. Taking after the Nord Lead and Access Virus line of analog modeling

Figure 8.14 Access Virus analog modeling synthesizer.

synthesizers, Roland has entered the analog modeling market with their Aira line of synthesizers. Roland's new Aira line uses what they call "Analog Circuit Behavior" technology. Roland's "Analog Circuit Technology" is really just analog modeling technology with a higher number of individual algorithms. Nonetheless, the Aira line sound amazing and respond as if they are truly analog synthesizers. The Aira line consists of the flagship *System-1 Synthesizer*, *TR-8 Rhythm Performer*, *TB-3 Touch Bassline*, and the *VT-3 Voice Transformer*. Each of these synthesis devices are extremely powerful and display an incredible use of analog modeling technology. Roland also introduced a new "plug out" technology into the *System-1* that allows users to perform a system dump of a

Figure 8.15 Roland Aira *System-1* analog circuit behavior synthesizer.

VST synthesizer into the *System-1*'s memory so that it can be recalled without having to be connected to a computer. The Aria line of synthesizers and most notably the *System-1* are an exciting direction for the analog modeling climate.

Workstation and Combination Synthesizers

As digital recording became more prevalent in the 1990s, synthesist's demands for synthesizers capable of being relevant in the digital recording studio became more widespread. As a response, companies like Roland, Yamaha, and Korg started producing *workstation* synthesizers. Workstation synthesizers are basically keyboards that feature a number of synthesis engines as well as multitrack recording and sampling capabilities. These workstation synthesizers will typically feature a number of synthesis engines such as subtractive, FM, sample based, etc. In addition to offering a number of sound synthesis engines, many workstation synthesizers will allow users to combine various elements of different synthesis engines. For example, a user could create a sound through FM synthesis, and then apply a resonant low pass filter such as is found on subtractive synthesizers. This ability to combine synthesis techniques is extremely attractive because it allows users to free themselves from the confines of a single synthesis format. The best attributes of each synthesis format can be combined in order to create one customized synthesizer that is perfectly suited for the user. *Combination synthesis*, as this combining functionality is termed, is found on most modern software synthesizers. Most software synthesizers will allow some sort of combination synthesis whether it is fully functional synthesis engines, or just a simple low pass filter.

The number of synthesizer workstations that have been released, and are subsequently on the new and used markets, could fill many chapters and do not warrant an in-depth model by model exploration. That being said, the Korg Kronos and Oasys are among the first workstation synthesizers and will serve to outline the features one might expect to find in a workstation and combination synthesizer.

Korg Kronos and Oasys

Both the Korg Kronos and the Oasys before it are powerful workstation synthesizers that boast nine discreet sound synthesis engines. The various synthesis engines available to Kronos and Oasys users include virtual subtractive synthesis, sample-based synthesis, physical modeling synthesis, wave shaping, and frequency modulation. The Oasys was Korg's flagship workstation synthesizer up until the Kronos was introduced with better memory and greater computing power.

Besides just allowing for complex synthesis capabilities, the Kronos allows users to digitally record up to 16 tracks at 24-bit resolution as well as full MIDI connectivity. The trend of including digital recording capabilities on synthesizers became extremely widespread in the early 2000s and is still very much found today. Although widespread use of DAW software has somewhat eclipsed the usefulness of recording capabilities on workstation synthesizers, many users still desire the function ability.

Workstation Breakdown

Although what is desired in a workstation synthesizer is constantly changing, a few key elements will most likely be present in most, if not all, workstation synthesizers. These elements include a number of individual synthesis engines, as well as a complex sampling interface, both a complex sequencer and arpeggiator, multitrack digital audio recording interface, full MIDI connectivity, and, finally, a wealth of controller types including keys, sliders, knobs, and drum pads. In today's ever-changing digital recording climate, the desire for workstation keyboards is dwindling. Multitype MIDI controllers seem to be replacing workstation synthesizers since they can easily be utilized in a DAW environment such as Logic Pro, FL Studio, or Ableton Live. Perhaps in an ironic twist of fate, all-in-one workstation synthesizers will regain popularity in the future and be coveted as "vintage synthesizers."

Recipes

The recipes provided in this chapter will be geared towards physical modeling synthesis as it is the only completely unique synthesis format introduced in this chapter. For many years, the allure of modeling synthesis was its uncanny ability to re-create sounds with the utmost precision. Many people felt that this was modeling synthesis's most prominent strength. Although modeling synthesis's ability to re-create the sounds of physical instruments is astounding, we feel its true strength is its ability to create new sounds not attainable through any other form of synthesis. Therefore, we have chosen to take this direction in the creation of the recipes for this chapter. We chose to use the modeling soft synth Sculpture, which comes with the DAW Logic Pro. Before we delve into each of the recipes, let's take some time to explore, in more depth, what Sculpture has to offer.

Sculpture

Sculpture is an immensely powerful modeling synthesizer. Besides its awe-inspiring modeling engine, Sculpture features a wealth of resonant filters, wave shaping capabilities, powerful DSP effects, and an immensely cool vector-inspired movement section. Add to this the powerful modulation matrix and various performance control features and Sculpture shines as a software synthesizer. To stay true to the modeling aspect of this chapter, we have chosen to create patches that only use Sculpture's modeling engine. Therefore, traditional filters and effects will not be found in these recipes. Instead, the ten recipes that follow will serve to explore the sheer power that modeling synthesis has to offer. Before going through each of the recipes in depth, however, it is necessary to explain a few of the parameters we will be using inside of Sculpture.

Figure 8.16 Sculpture's string section.

The String

Sculpture's sound engine is based around the concept of a string. The entire synth is designed so that the user can determine the material this string is made of, how the string is excited into motion, how the string's sound is captured, and then, finally, how the captured sound is modulated and effected. The circular section in the middle of the interface is where we can determine the material and stiffness of the string. Inside this circular section is a square labeled *Material*. By moving the sliver ball around this material square, we can determine the material the string is made up of. Our options are nylon, wood, steel, and glass. It is important to note, however, that similarly to a vector plane, this ball can be moved anywhere in this square allowing for an extreme amount of tonal changes.

Objects

Once we have determined the material of our string, we can choose up to three objects to excite said string. Each of the three objects has a wide variety of *types*, which relates

Figure 8.17 Sculpture's object section.

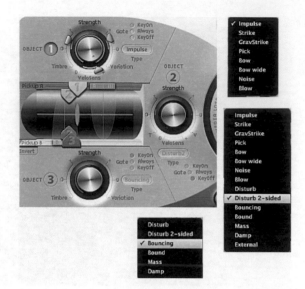

to how the string will be excited. The range of types is vast but includes thing like bouncing, bow, strike, and impulse. All of these types are models. This means that if one were to choose the *bouncing* object type, Sculpture would produce an algorithm that mimicked something physically bouncing on the main string. The sounds that can be created with these models are truly breathtaking. You can think of these objects as oscillators with the *type* selector akin to a wave shape selector. Each object and *type* will add sonic character to the sound as a whole. We also have the ability to choose how the object responds to a key being pressed. We can choose to have the object excite the string either when a key is depressed, released, or have the object constantly exciting the string. A few additional options are available to us in which to further sculpt how the object will excite the string.

Pickups and Object Positions

The Pickup section of Sculpture allows us to choose how we capture the sound emanating from the main string. Think of the two Pickups like pickups on a guitar, based off where the pickups are on a guitar's body will determine the timbre of

Figure 8.18 Sculpture's Pickup and object position section.

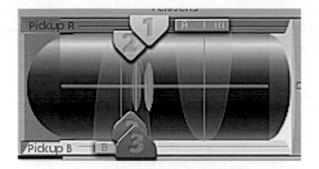

the sound. The two Pickups in Sculpture, *A* and *B*, can be separately moved across the entire length of the string. The two Pickup positions can also be modulated. Inside the Pickup section, we can find the object positions. These controls allow us to move each of the three objects to different areas of the string, causing different responses and timbres. Each of the three objects can be placed at any point on the string

LFOs and Modulation

Sculpture offers two freely assignable LFOs that can be routed to virtually any parameter on the synth, a third LFO

Figure 8.19 Sculpture's modulation section.

dedicated to pitch modulation, or vibrato, two envelope generators, a jitter generator, and, finally, two randomizers. Each of these parameters can be sculpted by the user and routed to a variety of places for extreme modulation capabilities. The two main LFOs have a wealth of wave shapes and can be free running or beat synced.

Master Section

The final parameter in Sculpture's modeling engine is the *Master* section. In the *Master* section, we have the ability to adjust the main amplifier envelope generator as well as the panning of keys and Pickups, and then, finally, the overall level of the synth.

Other Parameters

As stated earlier, we are just focusing on the modeling engine inside of Sculpture, but there are a wealth of other parameters that Sculpture offers that warrant mention even though we will be disregarding them in our recipes. First and foremost is Sculpture's powerful multifilter. Sculpture's filter is fully resonant with high pass, low pass, peak, band pass, and notch filter shapes. Next is the wave shaper. Adjusting Sculpture's wave shaper will add an amount of distortion or harshness to a patch that is quite astounding. Finally, a full-fledged morph pad and corresponding morph envelope allow for extreme amounts of animated movement, which is sure to take your patch to the next level.

Now that we have covered the basics of Sculpture's modeling engine, let's explore each of the patches we have created for

Figure 8.20 Sculpture's *Master* section.

this chapter. If you own Sculpture, feel free to follow along and expand upon these patches.

Recipe 1: Stiff Bouncing Shimmer

The first recipe on our list sounds like a soft mallet bouncing on the high strings of a piano. The sound is quite pretty and would fit well in most pop or ambient songs. The patch consists of all three objects with a string material that is between steel and nylon with about 60% resolution. Object one is set to *Impulse* with a *keyOn* gate. Object two is set to *Bow* with a *KeyOn* gate as well. Object three is set to *Bouncing* with a *KeyOff* gate so that the bouncing sound comes in once a key is released. Object one has about 75% strength while object two has 50% strength. Object three is weaker still, with about 20% strength. All the objects are placed in chronological order on the string with object one starting about 1/5th up the string from the left-hand side, while object three is about 4/6ths up the string. The amplifier's envelope has zero attack with medium range decay and release and a slightly lower sustain. Both *Key* and *Pickup* are panned to the center. Pickup A is set at 1/4th up the string from the left while Pickup B is set at 3/4ths up the string from the left. A sine wave LFO set to 0.75Hz is routed to Pickup B's pan control, almost fully engaged in the positive domain.

Figure 8.21 Screenshot of Stiff Bouncing Shimmer patch.

Figure 8.22 Screenshot of Distorted Dissonant Pop patch.

Recipe 2: Distorted Dissonant Pop

At first, this second recipe sounds like a distorted FM synth sound, but then it quickly turns dissonant in an extremely pleasing manner. The patch consists of just the first two objects both turned to *Impulse*. Object one has a strength setting of 100% while object two is set to 50% strength. The string material is fairly close to the *Steel* corner with full resolution. Object one is placed about 1/3rd up the string from the left, while object two is placed directly in the center. Pickup A is placed about 1/4th up the string from the left, while Pickup B is placed 3/4ths up the string. The *Pickup* parameter is spread modestly into the left and right while *Key* is fully spread. The amplifier envelope is set to an extremely short attack, midrange decay, full sustain, and nonexistent release. Finally, a 100Hz sine wave LFO is routed to the position of both Pickups with about 30% depth in the positive domain.

Recipe 3: Realistic Muted Bass

The third recipe on our list is extremely reminiscent of a slightly muted electric bass being played with a pick. The sound consists of the first two objects with a type setting of *Impulse* and *Strike*, respectively. Both objects feature about a 50% strength setting. The string material is set between steel and nylon with the setting close to nylon and about 60%

Figure 8.23 Screen shot of Realistic Muted Bass patch.

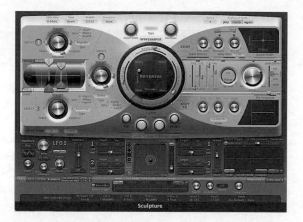

resolution. Pickup A is placed about 1/5th up the string from the left while Pickup B is placed about 3/5ths up the string. Object one and two are placed on either side of the halfway mark with about a full 1/4th space between them. The amplifier envelope features an extremely short attack time, midrange decay time, full sustain, and about a 3/4ths full release time. The *Pickup* setting is fully spread.

Recipe 4: Odd Shimmering Pad

The fourth recipe on our list is a weird pad-type sound that can only be explained by imagining a steel bow scraping

Figure 8.24 Screenshot of Odd Shimmering Pad patch.

across a resonant steel sheet. The sound consists of all three objects with *Strike*, *Blow*, and *Bouncing* type settings respectively. Object one has a 50% strength setting, while objects two and three have a 25% setting. Object one is placed 3/4ths up the string from the left, while object two is placed 1/4ths up and object three placed about halfway up. Pickup A is in line with object two while Pickup B is placed directly before object one. The string material is set between steel and glass with the selector being closer to steel with full resolution. The amplifier envelope features a 3/4ths-full attack time, half-full decay, full sustain, and 3/4ths-full release. Both *Key* and *Pickup* are fully spread. A 6.70Hz sine wave LFO is routed to the panning of Pickup A and B with 100% positive depth, while a second 0.08 sine wave LFO is routed just to the panning of Pickup A with about a 50% positive depth.

Recipe 5: Deep Synth Bass

The fifth recipe on our list is reminiscent of a traditional, eighties' synth bass, but with a weird elastic rubber quality to it. The sound is made up of all three objects with *Strike*, *Pick*, and *Bouncing* settings respectively. Object one has a 100% strength setting, while both object two and three feature 50% strength settings. Object one is placed just shy of 3/4ths up the string from the left, while object two is placed 1/5th up and object three is almost directly between objects one

Figure 8.25 Screenshot of Deep Synth Bass patch.

and two. Pickup A is perfectly in line with object two, while Pickup B is perfectly in line with object one. The string material is between nylon and steel with heavy emphasis towards steel and full resolution. The amplifier envelope features an extremely short attack with a midrange decay, full sustain, and 3/4-full release.

Recipe 6: Vibrating Bells

The sixth recipe on our list sounds a bit like if one were to gently strike bells with a felted mallet. The sound consists of the first two objects set to *Impulse* and *Bow* respectively. Object one features a strength setting of about 90%, while object two's strength is set to about 75%. Object one is placed about 1/5th up the string from the left, while object two is placed 4/5ths up. Pickup A is perfectly in line with object one, while Pickup B is placed about 3/5ths up the string. The string material is between steel and glass but inching towards the middle of the plane, allowing for a slight mix between all the materials. The string features full resolution. The amplifier envelope features an extremely short attack, midrange decay, full sustain, and 3/4-full release. *Key* is slightly spread. A 0.19Hz sine wave LFO is routed to object two's position with 100% positive depth. An additional 0.07Hz sine wave LFO is routed to the position of Pickups A and B with 100% positive depth.

Figure 8.26 Screenshot of Vibrating Bells patch.

Figure 8.27 Screenshot of Full Spectrum Synth patch.

Recipe 7: Full Spectrum Synth

This seventh recipe is a cool synthesized mix between a low guitar string pluck and high-end piano. It's made up of just two objects with *Impulse* and *Noise* type settings, respectively. Object one is set to a *KeyOn* gate with 100% strength, while object two is set to a *KeyOff* gate with about 65% strength. Object one and two are both placed about 1/3rd up the string from the left with object one slightly further to the right than object two. Pickup B is slightly to the left of object two while Pickup A is about 3/4ths up the string from the left. The string material is in the bottom left corner, allowing for a mix between steel, glass, and nylon with most emphasis placed on steel. The string features full resolution. The amplifier envelope features an extremely short attack, almost full decay, zero sustain, and slightly more than half of a full release. Finally, a 0.12Hz triangle wave LFO is routed to the position of Pickups A and B with about 36% positive depth.

Recipe 8: High Rate of Wobble

The eighth recipe on our list sounds like an acoustic guitar being strummed under water. It's made up of all three objects with *Impulse*, *Noise*, and *Bouncing* type settings, respectively.

Figure 8.28 Screenshot of High Rate of Wobble patch.

Object one features 100% strength setting, object two features 34% strength setting, and object three features 62% strength setting. Objects one and three are placed almost on top of one another about 1/3rd up the string from the left, while object two is placed 3/4ths up the string from the left. Pickup A is placed directly to the left of objects one and three, while Pickup B is placed directly to the left of object two. The string material is in the bottom left of the square resulting in a mixture between nylon, steel, and glass, with a heavy emphasis on steel. The string features full resolution. The amplifier envelope generator features an extremely short attack with almost full decay, nonexistent sustain, and a little more than half of a full release. The *Pickup* parameter is fully spread. A 31Hz triangle wave LFO is routed to the panning of Pickups A and B with 100% positive depth, which results in the underwater aspect of the sound.

Recipe 9: Synth Keys

The ninth recipe on our list is reminiscent of an electric piano: only with a metallic, synthetic flair. The patch consists of only the first object engaged with a *GruvStrike* type setting and a 75% strength setting with an *Always* gate. Object one is placed directly in the center of the string with Pickups A and B 1/4th in from either side of the string. The string material is in

Figure 8.29 Screenshot of Synth Keys patch.

between nylon and steel, but much closer to steel. The selector ball for the string material is a bit inward in the plane, so a mix is able to be heard of all materials. The string also features a resolution of 100%. The amplifier envelope features an extremely short attack, full decay and sustain, and, finally, a 1/4th-full release. *Pickup* is fully spread. A 1/2-beat sine wave LFO is routed to the position of Pickups A and B with a 20% negative depth.

Recipe 10: Distant Rhythms

The final recipe on our list is one of the coolest. It's a highly animated pad that sounds like an electronic music concert heard from outside of a tent. This patch was designed to be used in the lower register of a keyboard. This patch consists of only the first object engaged with a *Noise* type setting and about 50% strength. The timbre and variation settings on the object are both set at 100% in the positive direction. Object one is placed slightly to the left of center while Pickups A and B are both 1/6th in from either side of the string. The string material is almost fully set to steel with a 76% resolution. The amplifier attack is extremely short, while both the decay and sustain parameters are set to full. Release is set just shy of halfway. The *Pickup* parameter is almost fully spread. A 1/4 beat "Rect01" shaped LFO is routed to the positions of

Figure 8.30 Screenshot of Distant Rhythms patch.

Pickups A and B as well as object one's timbre. The routing to the Pickups has a 100% positive depth, while the object timbre routing has a 100% negative depth. This first LFO also features a 48% positive curve setting. The second LFO is a 1/8th-beat sawtooth LFO with a negative 16% curve setting. This LFO is routed to object one's timbre as well, only with a 40% positive depth.

Historical Perspective on Modeling Synthesis

Like some of the other synthesis formats discussed in this book, physical modeling synthesis was conceived before the technology existed with which to implement it. The first strides in physical modeling synthesis came from Lejaren Hiller and Pierre Ruiz. Hiller and Ruiz, who worked out of the State University of New York and Bell Laboratories, respectively, conceived of a way of creating realistic models of physical instruments through using finite difference approximations of the wave equation.[1] In essence, Hiller and Ruiz examined vibrating objects, such as strings, and attempted to describe them via differential equations that could then be coded and programmed for a digital computer. Although Hiller and Ruiz only concentrated on vibrating strings, they were able to prove their methods for use with other vibrating objects such as bars, plates, membranes, and spheres. Although Hiller and Ruiz were able to successfully model limited stringed instruments, it was not until the development of the Karplus-Strong algorithm that physical modeling synthesis started to gain traction in the scientific world.[2]

Alexander Strong invented, while Kevin Karplus analyzed and described, the algorithm that would become known as the Karplus-Strong algorithm. The algorithm models the sound of a plucked or hammered string by looping a short waveform through a filtered delay line. Imagine a plucked string, such as a cello played in a pizzicato manner. Directly after the string is plucked, the sound will be harmonically rich. Shortly after the string is first plucked, however, the higher harmonics will start to drop in amplitude followed by the lower harmonics, until the sound is finally inaudible. When using the Karplus-Strong algorithm, a short waveform is produced and then looped back into a low pass filter, eliminating some of the higher frequencies. The new, filtered waveform is looped back into the filter and more high harmonics are filtered out of the sound. The resulting waveforms are continuously looped and filtered until the sound is inaudible, mimicking the way in which a plucked string would naturally fade away. Each loop corresponds to approximately one period, or cycle, of the waveform.

The introduction and success of the Karplus-Strong algorithm piqued the interest of a lot of engineers and synthesizer companies. Brief murmurings about the promise of this new technology could be heard among a few, select individuals. One such person was Julius Smith, who went on to create an extension of the Karplus-Strong algorithm, which would become known as digital waveguide synthesis.[3] Digital waveguides are basically computational models for the way in which acoustic waves move through physical media. Using waveguide synthesis, engineers were able to successfully model a number of instruments with extreme accuracy.

Although the theory was there, DSP power simply was not. It would not be long, however, until DSP power would catch up with physical modeling theory and begin being implemented in commercial instruments. Yamaha was one of the first companies to see the potential of physical modeling synthesis. In 1989, Yamaha signed a contract with Stanford University to begin designing instruments that would feature this new technology. The first such instrument to utilize physical modeling synthesis was the Yamaha VL-1 released in 1994. The VL-1 was a revolutionary instrument that held extreme promise. Although many people were able to realize the sheer power of VL-1, it was far out of the reach of many people's budgets. In order to try and bring more appeal to Yamaha's new synthesis technology, the company released the VL-7, which was a slightly scaled back but much less expensive version of the VL-1. The VL-7 proved to be quite successful and even encouraged more manufacturers to release their own physical modeling synthesizers. However, sampling technology and low budget rompler synthesizers overshadowed the much more powerful physical modeling synthesizers of the day, resulting in a declining number of sales.

Physical modeling technology would, more or less, stay dormant throughout the 1990s and early 2000s. Although the occasional physical modeling sound engine would show up in a workstation synthesizer every now and again, no worthwhile advancements were made in the technology. This all started to change with the coming of digital audio workstations and software-based synthesis engines. CPU power began to reach levels that allowed for extremely powerful synthesis engines without much computational strains. Physical modeling synthesis has since flourished in the software domain with programs such as Reason and Ableton Live offering users a wealth of physical modeling capabilities.

Physical modeling synthesis is often confused with sample-based synthesis and is often overlooked and disregarded. Because of the extreme realism that can be achieved using physical modeling synthesis, many people just assume it to be interchangeable with sample-based synthesis. This viewpoint, however, fails to realize the power physical modeling synthesis holds as an independent synthesis engine. The truth is that physical modeling synthesis can not only re-create physical instruments beautifully, it can create new and interesting sounds that do not exist outside the world of physical modeling synthesis. Because of this, physical modeling synthesis should never be overlooked and should, instead, be realized for what it is, an amazing and truly inspiring synthesis format.

Notes

1. Diana Deutsch, *The Psychology of Music*, San Diego, CA: Gulf Professional Publishing, 1999, pp. 132.
2. J. O. Smith, "Delay Lines," in *Physical Audio Signal Processing*, http://ccrma.stanford.edu/~jos/pasp/Delay_Lines.html, online book, 2010 edition.
3. Julius O. Smith, III, "Physical Modeling using Digital Waveguides," *Computer Music Journal* vol. 16, no. 4 (Winter 1992), pp. 74–91.

PERFORMING LIVE 9

Musicians have long realized the benefits of incorporating synthesizers into their performances. Besides just allowing for a richer sonic palette, the synthesizer can add a cool factor to a performance that is unattainable through any other means. Seeing someone on stage surrounded by stacks of keyboards and synthesizer panels reminiscent of a mad scientist in a dark laboratory is enough to kick a viewer's imagination into high gear.

Being that there are so many synthesizers and controllers currently on the market, there are virtually hundreds, if not thousands, of synthesizer setups that could be examined for live performance. Due to this, we will examine a few traditional setups, as well as various categories of synthesizers and controllers that one may use live. The categories of synthesizers and setups covered in this chapter will vary from all analog gear, to all digital gear, to all software gear, and, finally, a number of combinations of each. The pros, cons, connections and playability, and portability of each setup will all be considered.

Using Analog Synthesizers Live

Performing live with analog synthesizers is not for the faint of heart. Analog synthesizers are finicky at best and downright unreliable at worst. In the very best of cases, a synthesist must make an extreme amount of control changes during a song as well as the short time between songs to create various sounds. More often than not, however, synthesists must constantly be tuning and maintaining their analog synthesizers throughout a single performance as well as troubleshooting circuitry problems mid show. That being said, analog

Figure 9.1 Classic live setup featuring analog gear.

synthesizers are capable of some of the warmest, richest, most mind-blowing sounds imaginable. Because of this, many musicians are determined to use them live despite the headache they may cause. Before delving into ways in which one may utilize analog synthesizers in a live environment, let's examine in depth some of the pros and cons of using analog synthesizers live.

Pro—Extreme Sonic Manipulation

Due to the nature of analog synthesizers having designated controls for every single parameter available, a musician is free to adjust these controls on the fly. Having control over the entire synth without having to menu dive on a small LCD screen allows the user to make changes with ease to virtually any parameter.

Pro—Logical Interfacing with Musicians

Most analog synthesizers require the synthesist to tune the individual oscillators to a set pitch. While this may seem like a chore, it actually allows for much easier collaborations between musicians and synthesizers. Take an orchestra for example. When musicians are tuning their respective instruments, they tune to one another rather than a pure tone. By doing this, the orchestra as a whole might be slightly off of A 440. A synthesist can then adjust the tuning of his or her analog synthesizer in order to be perfectly in pitch with the orchestra. The same theory holds true in a rock band where

Figure 9.2 Visual representation of an envelope follower.

Audio input signal

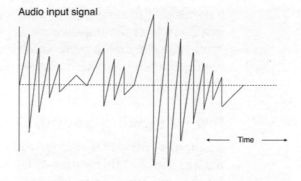

Envelope output signal

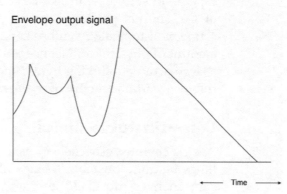

the synthesist can tune to the guitar or bass player. In addition to this, many analog synthesizers will feature a circuit known as an *envelope follower*. An envelope follower basically applies the envelope of a signal, such as a guitar, voice, or drum set, and then applies it to the synthesizer. While digital and software synthesizers surely offer this capability as well, analog synthesizers will typically allow for more fine tuning and customization of the envelope follower in order to ensure the synthesist gets the sound he or she desires.

Pro—Sonic Improvisation

As previously mentioned, analog synthesizers oftentimes feature designated controls for each and every parameter. Because of this, a synthesist is free to constantly change the timbre of the instrument. Having such unlimited control over the sound of the synthesizer allows the synthesist to keep up in a heavily laden improvisational setting. For instance, the synthesist can create more subtle sounds as the band plays more *piano* and harsher sounds when the band plays more *forte*.

Pro—Amazing Sound/Cool Factor

It is hard to deny that analog synthesizers are capable of producing some of the warmest, fattest, most inspired sounds capable in sound synthesis. Besides the sheer awesomeness of the sounds produced, many synthesists feel a certain connection to analog synthesizers, perhaps due to the large amount of physical contact needed to perform on one. Lastly, a large array of blinking lights, knobs, and even patch cords convey a certain amount of coolness that many people desire.

Con—Unstable Circuitry

Analog circuitry, especially vintage analog circuitry, can oftentimes be quite unstable. Things like temperature and pressure changes, dust, old age, and fluctuating voltages can wreak havoc on an analog synthesizer. At best, the user will experience oscillators that drift in and out of pitch and must be recalibrated. At worst however, analog synthesizers can begin to crackle, pop, and even suddenly cease making noise.

Con—Patch Storage/Performance Mapping

For many years, analog synthesizers simply did not feature patch storage. Once they started to incorporate patch storage in the late 1970s, it was extremely limited. Although many modern analog synthesizers will offer patch storage, a good percentage of the analog synthesizers being used do not. Another function found on many digital synthesizers, but absent on analog synthesizers, is performance mapping. Performance mapping is designating individual preset sounds to different ranges of the keyboard in order for the synthesist to have access to different sounds at once. Analog synthesizers are almost wholly incapable of this function.

Con—Large, Cumbersome, and Limited

Analog synthesizers are oftentimes seen as large, heavy machines. Although a MiniMoog is leaps and bounds smaller than its Moog modular counterpart, it is still deemed large in today's musical environment. Therefore, many users feel that it is a burden to lug around analog synthesizers from show to show. Furthermore, many musicians choose to use multiple analog synthesizers on stage at once to have different sound options. Bringing multiple analog synthesizers to every show can start to take a toll on a musician who does not have a large bus and roadies.

A Note on Unstable Circuitry

Although all analog circuitry is susceptible to environmental changes, most modern analog synthesizers are much more stable than their vintage counterparts. In fact, once a modern analog synthesizer has fully warmed up (typically 20–30 minutes after powering on), it will most likely stay in tune for the duration of the show. Although modern analog synthesizers

Figure 9.3 The MiniMoog Voyager is a modern analog synthesizer.

are extremely stable, many musicians choose to perform with vintage models, so unstable circuitry must remain as one of the potential cons of using analog synthesizers live.

Controlling Analog Synthesizers Live

Once the decision has been made to use analog synthesizers live, a user must decide how to control his or her analog synthesizers. Luckily, there is a wide variety of controllers and setups available to the analog synthesizer purist. Before delving into the various controllers available, a discussion on control voltage is warranted. Control voltage, or simply C.V., is used to determine pitch and gate information on an analog synthesizer. Because analog oscillators produce a continuous pitch, something is needed in which to map out various notes so the synthesizer can be played musically. Traditionally, synthesizers use the 1volt/octave standard, meaning one volt is divided into 12 equal parts to accommodate the Western 12-tone scale. This means that two "A" notes played an octave apart will also be one volt apart. Not all synthesizers adhere to this volt/octave standard, though, so special precaution must be taken when interfacing multiple analog synthesizers. Control voltage is also used for creating gate signals on an analog synthesizer. Because an analog oscillator is always producing sound, an amplifier and gate trigger are needed in order to make sound heard only when a key is depressed. Traditionally, this voltage is +5volts.

As you probably already know, many synthesizers feature a piano-like black and white keyboard, which acts as the synthesizer's pitch and gate controller. These keyboards will send control voltage to the synthesizer whenever a key is depressed or let go. The easiest way in which to play an analog synthesizer live is to simply set up a few synths and play away on the key beds. When using this approach however, a synthesist is limited to only playing two synths at once with one hand on each keyboard. Although this surely works and has been used by thousands of musicians, there are many instances when an additional synthesizer is desired to play a droning note or a pattern that changes pitch with the song. By using the control voltage outputs on an analog synthesizer, one can control a variety of synthesizers with just a single keyboard.

Patching the control voltage and gate outputs of a synthesizer into the control voltage and gate inputs on an additional synthesizer will allow for simultaneous control. When used in this way, a synthesist is able to trigger the second synthesizer via the first synthesizer's keyboard. This technique is extremely useful when creating extremely large synth sounds that are far too big to be contained on a single synthesizer. A similar and equally awesome technique is to only patch the pitch control of the first synthesizer into the second synth while opening the amplifier of the second synth. By doing this, a pattern can be played on the first synth while the second synth produces long, droning notes that match the pitches played on the first synthesizer. When using this technique, there are a variety of ways in which to open the filter of the second synthesizer that are dependent on the synthesizer itself. Some synthesizers feature a button that leaves the filter continuously open; this is usually labeled something like "hold," while other synthesizers will require an external continuous gate source to be patched into them.

While these techniques are extremely useful and quite musical, some musicians require additional patterns and synth lines that cannot be created using just two hands. One way in which to produce these additional patterns and lines is to use a laptop or playback device. Although laptops and playback devices certainly have their place, this discussion is centered on analog gear, so we will focus for the time being on sequencers and arpeggiators.

Sequencers have long been a staple in synthesis. A sequencer is basically a device which spits out pitch and gate information at regular intervals. Sequencers can be used in a number of ways, but by far the most traditional use of a sequencer is a repeating pattern that stays constant or gets transposed

Figure 9.4 Multiple synthesizers connected through control voltage. Synthesizer photos courtesy of www.switchedonaustin.com.

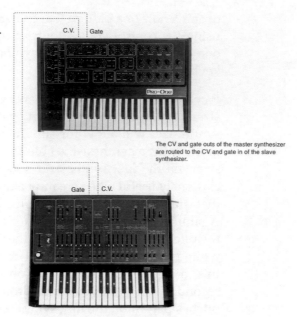

The CV and gate outs of the master synthesizer are routed to the CV and gate in of the slave synthesizer.

with key changes. As sequencers only produce control voltage and gate information, they can be used to create a pattern on any synthesizer parameter that accepts control voltage, such as a filter. By patching a sequencer into a filter, the filter will open and close rhythmically based off of the settings of the sequencer. Pete Townshend is famous for employing this technique on a number of Who songs, the most notable being "Baba O'Riley."

A sequencer gets its timing information from a gate source known as a clock, which is usually a pulse wave-producing oscillator. Most sequencers will feature internal clock sources as well as have the ability to be clocked from an external source, such as a separate synthesizer or metronome. Using a master metronome or clock source for a number of different synths and sequencers will allow for all synth patterns to be played in time with one another, which will help the analog synthesist in performance. By using a sequencer in this way in conjunction with other synthesizers, complex patterns can be created while the performer is free to play lead lines over the pattern.

Like a sequencer, an arpeggiator is a device that also produces pitch and gate information in a repeating manner. An arpeggiator will almost always be built into a given synthesizer and is more limited in scope than a sequencer. An arpeggiator gets its pitch information from keys that are physically depressed on the synth's keyboard and repeats them in low to high or high to low order. Many arpeggiators allow for the addition of octaves of the notes being played. Some arpeggiators even allow for dotted or swung rhythms. Unlike sequencers however, arpeggiators almost always get their clock from an internal source and typically don't allow for external clocking. Despite their more limited nature, arpeggiators are great for creating simple, repeating patterns that bode well to the live performance environment.

Using Modular Synthesizers Live

In recent years, a resurgence of analog modular synthesizers has taken place with a number of companies creating new modular gear in both traditional 5U and newer Eurorack sizes. The modular synthesizers of today are in many

Figure 9.5 A clock source is used to send timing information to various devices.

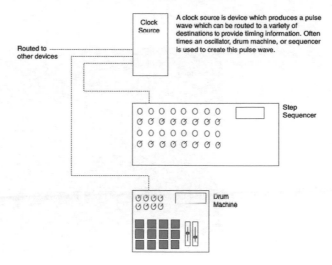

aspects more powerful and even better than modular synthesizers created in their heyday in the 1960s and 1970s. Analog circuitry has come a far way since the 1960s and allows for much more stable synth modules. While not many people even toyed with the notion of bringing their modular synthesizers onto the stage in the 1970s, many musicians are beginning to perform with modular equipment.

A modular synthesizer is made up of individual modules that each act as standalone synthesizer blocks. This means that in order to have a three-oscillator, two-envelope generator, and single-filter modular synthesizer, three oscillator modules, two envelope generator modules, and a filter module must all be present in a system. The number of individual modules a system contains is typically left up to the user as he or she can purchase and add on as many modules as he or she wishes. Being that each module exists separately from one another, patch cords must be connected between modules. Using modular synthesizers live requires the same care given to analog synthesizers as well as a wealth of spare patch cables and parts. Using a modular synthesizer live can be considered the ultimate creative tool for sonic manipulation, but like with anything else, it takes practice to perform with a modular synthesizer.

Figure 9.6 A modern, modular synthesizer setup.

You are probably starting to notice that creating interesting, complex analog synthesis performances requires a wealth of equipment all behaving nicely with one another. In order to create an environment with evolving patterns and additional lead lines, the minimum equipment needed is a sequencer, a synth that will play said sequence, and an additional synth to play lines of the pattern. If more than one synth sound is desired in a song, additional synths will have to be added to the mix. Because analog circuitry is finicky, constant tweaking will be necessary not only to keep the synths working and communicating correctly with one another, but also to maintain stable tuning and creating various sounds between songs. That being said, this is a small price to pay for the benefit analog synthesis has to offer.

Using Digital Synthesizers Live

If lugging around a large number of synthesizers and constantly tweaking them to maintain them does not seem like your cup of tea, digital synthesis may be right for you. In fact, digital synthesizers became popular for this very reason; musicians were sick of bringing so many synthesizers on stage just to have a variety of different synth sounds. Once digital synthesizers like the Yamaha DX7 started to hit the market, musicians were finally free to bring a single synthesizer on stage that was capable of switching sounds in an instant while maintaining extreme levels of stability. The Yamaha DX7 and digital synthesizers like it helped all but kill analog synthesis until recent years. As with analog synthesizers, let's explore some of the pros and cons of using digital synthesizers live before we delve into some of the possible setups a musician can use.

Figure 9.7 The Yamaha DX7. Photo courtesy of www. switchedonaustin.com.

Pro—Preset and Patch Storage

Perhaps one of the biggest advancements that digital synthesizers brought to live performance was the ability to not only to change the synthesizer's sound instantaneously, but also to store patches as presets. By allowing patch storage and instantaneous recall, musicians were free to start a song with a pad sound and then change to a screaming lead sound in an instant without changing any settings on the synth itself. Many musicians flocked to this technology with just the promise of preset technology. Digital synthesizers such as the Yamaha DX7 also allowed musicians the capability of re-creating natural instrument sounds such as organs and electric pianos very easily, which allowed them to leave their heavy organs and electric pianos at home when they went on tour.

Pro—Extremely Stable Circuitry

Due to the presence of integrated circuits and digital signal processing (DSP) technology in digital synthesis, digital synthesizers are extremely stable machines. Unlike analog synthesizers, which are susceptible to environmental changes, digital synthesizers will typically remain 100% stable from powering up to powering down. Being able to fully rely on a synthesizer through not only the duration of a show but an entire tour was and is hugely attractive to most musicians.

Pro—Small, Compact Size

Although analog synthesizers began getting smaller and smaller before digital synthesizers took over, most were still too large and cumbersome for many musicians. Digital synthesizers, on the other hand, were often much more compact than their analog counterparts. Offering the powerful tools of digital synthesis in a relatively small, compact package not only allows for ease of portability, but also allows musicians to have more tools and sounds at their disposal than could be had with an equal number of analog synthesizers.

Con—Less Instantaneous Customization

Although digital synthesizers allow for instantaneous preset recall, customizing each preset on the fly is troublesome. Many digital synthesizers require users to dive deep into the many menus and submenus of the synth itself in order to adjust various parameters such as filter cutoff, resonance, and envelope generator settings. Although many musicians might feel that simply recalling and playing presets is adequate for their performances, many people prefer the ability to adjust parameters live. That being said, many modern digital synthesizers have included the ability to make real-time adjustments to various parameters with the inclusion of potentiometers and sliders dedicated to each parameter.

Con—Limited Field Repair

Although no musician hopes for their synthesizer to suddenly stop working on a tour, it oftentimes cannot be avoided. When touring with analog equipment, most repairs are fairly simple to someone with even a basic understanding of electronics. Parts can be purchased and swapped out on an analog synthesizer with only a few hours and a soldering iron. Digital synthesizers, although more stable than analog, seem to fail more catastrophically. It should be known that most digital synthesizers, especially modern digital synthesizers, will rarely fail, but if they do, they might not be repairable on tour and would have to be shipped to the factory. If a digital synthesizer suddenly stops producing sound, it might be a more complex fix than simply tracing signal through the synth or swapping out the output jack. In these instances, a complete overhaul of the internal memory, DSP, or motherboard might be necessary.

Con—Less Flexibility When Interfacing with Musicians

Although analog synthesizers are finicky, they are more suited for direct interaction with other musicians in a live performance setting. Digital synthesizers mainly require MIDI information or physical playing to control them. This means that in order to be triggered from a drummer, the drummer must be using either a MIDI drum set or have access to MIDI pads. This is not to say that musicians have not incorporated inventive ways to get around this and successfully integrated themselves with digital synthesizers, only that digital synthesizers are not as easily suited for this purpose as analog synthesizers.

Controlling Digital Synthesizers Live

Like with analog synthesizers and control voltage, digital synthesizers have their own communication protocol known as Musical Instrument Digital Interface, or simply MIDI. MIDI is a proprietary control technology that is used in many areas outside of synthesizers and even outside of music. MIDI has become the dominant control force in the industry since its inception in 1983. Virtually every synth created today, be it analog, digital, or software, will be able to send and receive MIDI messages. Basically, MIDI is a two-way communication protocol that relays information such as notes, velocity, controller wheel positions, etc. Using MIDI, musicians can connect various synthesizers in a similar fashion to their analog counterparts, connect designated controllers to MIDI sound racks, and, finally, connect synthesizers and controllers to a computer.

Like with analog synthesizers, digital synthesizers feature step sequencers and arpeggiators as well. As might be guessed, sequencers and arpeggiators that exist on digital synthesizers will most likely be completely digital. In the case of step sequencers, rather than sending out voltages for pitch and gate information, digital sequencers will output MIDI data, which will control the synthesizer. Digital sequencers and arpeggiators will most often be capable of longer, more complex, and more customizable patterns. Digital sequencers and arpeggiators can be used in the same way as they are used in the analog world.

One feature digital synthesizers offer that is rarely, if ever, seen on analog synthesizers is the ability to key map. Basically, key mapping refers to the ability to place individual presets onto different ranges of the synthesizer's keyboard. Using key mapping, a bass patch can be placed on the lowest two octaves of the keyboard, a pad sound on the middle octaves, and, finally, a lead patch on the highest octaves of the keyboard. Key mapping offers musicians the ability to have multiple sounds on a single keyboard that can be played together or separately without having to physically move to a different synthesizer. Effects that might be present in a digital synthesizer can also usually be key mapped allowing for evolving effects as one moves up and down the keyboard. When interfacing laptops, tablets, or other playback devices with digital synthesizers, key mapping can be used to trigger samples or even lighting cues. The possibilities of key mapping are literally endless.

In my mind, one of the largest benefits to digital synthesizers and, more poignantly, MIDI is the ability to utilize sound modules and controllers in a live set up. A sound module is a rack-mounted device that has a large number of presets and effects that can be controlled via a MIDI controller. Although this type of technology is not new and even exists in the analog world with devices like the Oberheim SEM modules, MIDI and digital technology allows for more possibilities than analog technology. By bringing a single MIDI sound module and a small MIDI controller, musicians are free to have as many sounds as they desire, ranging from strictly synthesized sounds to high quality samples. As we will explore later in the chapter, laptops and tablets have mostly replaced MIDI sound modules, but the theory and core concept of them stays the same.

Using Software Synthesizers Live

The climate of software synthesizers is constantly evolving. In the beginning, software synthesizers required tons of CPU power and were found only on desktop computers in studios. It was not until computer power, as well as the release of powerful laptops, began to increase that software synthesizers began appearing on stage. Today's software synthesizer climate looks different still with the use of hardware synthesizers, laptops, tablets, and even smartphones all running software synthesizers. As software synthesis is such a large topic, I will divide the discussion into various sections for each of the platforms one can run software synthesizers on for live performance.

Laptop Computers

Laptops are one of the most popular platforms with which to play software synthesizers live. Most laptops made in the last few years are fully capable of storing a ton of sound libraries and independent software synthesizers. With minimal equipment, just a controller, USB cable, and a laptop, musicians are free to have a whole arsenal of powerful synthesizers at their disposal. On average, most laptops will utilize MIDI over

Figure 9.8 Basic laptop and MIDI controller setup.

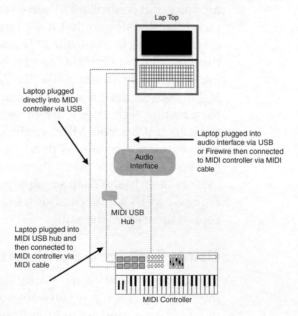

USB, meaning that MIDI information is transmitted to and from the computer via a USB cable rather than a traditional five-pin MIDI cable.

There are many benefits to performing live with a laptop besides just having access to powerful software synthesizers. The on-stage laptop can be running a DAW, or digital audio workstation, like Ableton Live, Pro Tools, or Logic, in order to playback samples and manipulate live audio in real time. Inside of the running DAW, the musician is free to have as many software synthesizers running as the computer or software can handle. Using a DAW live in this fashion allows musicians to bring the studio to the stage and perform in a way that might have seemed impossible a few decades ago.

Tablets and Smartphones

When the iPad was released, it changed the face of live performance for electronic musicians forever. Like with laptops, a multitude of software synths and workstations can be

Figure 9.9 Apple iPad.

utilized on stage inside a tablet or smartphone. This however, is where the similarities end between the two platforms. While laptops can be viewed as just glorified sound modules, tablets bring a new aspect to soft synth live performance that had been missing—physical interaction. Because tablets utilize advanced touch screen technology, app makers are free to incorporate all types of physical interaction scenarios into their programs.

Tablets can be used as either platforms for hosting software synthesizers and workstations or controllers with which to control software and hardware synths. To me, use as a controller is where the iPad, and tablets like it, really shine. Using MIDI over Wi-Fi or Bluetooth frees musicians from being physically connected to a laptop, sound module, or hardware synth, and allows them to move about the stage freely. Due to the advanced touch screen capabilities tablets have to offer, many exciting apps are available that allow for esoteric and interesting means of control. Some apps give users the ability to move around shapes, which act as parameter controllers, and bounce them across the screen with seemingly real physical properties, while some let you customize your own controller with as many knobs and sliders as you wish. Some apps of note that allow this type of control are "Touch OSC," "Control," "TouchDAW," and "MIDIDroide." The beauty about most of these tablet controller apps is that they are many times 100%-user customizable, meaning that you can designate a particular knob, slider, or shape to control anything that is transmittable via MIDI. Tablets seem to be bringing back the physical interaction between musicians and their synthesizers, which has been absent since analog synthesizers.

OSC

Shortly before the iPad came out, a different touch screen tablet device was beginning to gain huge traction among musicians and DJs. The device was called the Jazz Mutant Lemur and it attempted to be everything that music production on

Figure 9.10 The Jazz Mutant Lemur. Photo courtesy of www.switchedonaustin.com.

the iPad would eventually become. Although the iPad all but killed the physical Lemur (an app version of the Lemur exists and is extremely popular), the Lemur did change people's notions about how these types of devices could be used in a live setting. The Lemur is perhaps best remembered for bringing OSC (Open Sound Control) out of universities and into consumer's hands.

OSC is a protocol that was designed to be an alternative to MIDI. One of the benefits of OSC is its speed. Because OSC is transmitted through ethernet cables, it allows for immensely faster speeds than is possible through traditional five-pin MIDI cables. OSC also allows for much more customization on the part of the user than MIDI traditionally allows. Although it showed much promise, OSC seems to have drifted back out of the spotlight. Although it's no longer in the forefront, however, OSC did address problems with the MIDI protocol as well as offer new and inventive control means that are beginning to be incorporated into tablet MIDI controller apps.

A Note on Hardware/Software Synthesizers

Software synthesis is not limited to just laptops, tablets, and smartphones. On the contrary, many "hardware" synthesizers

Figure 9.11 Roland Aira System-1 synthesizer. Photo courtesy of www.switchedonaustin.com.

feature software synthesis sound engines. Synths like the Access Virus line, Roland Aria line, and Arturia Origin all create sound through software synthesis. Although these synthesizers surely seem like physical hardware synths, they are really just a MIDI controller, computer, and software built into a tidy box. These synths might not offer the flexibility of a standalone laptop and controller, but they have the ability to exist outside the realm of computer integration and are more comfortable to some musicians.

Controlling Software Synthesizers

Like with all other forms of synthesizers, software synthesizers feature unique ways in which a musician can control them in a live setting. As was discussed in the laptop section, the most common control setup is simply a MIDI controller connected via USB to a laptop running a software synthesizer. Although this setup seems fairly straightforward, a huge variety of MIDI controllers exist that should be examined.

A MIDI controller is any device that transmits MIDI data when played. MIDI controllers come in a variety of shapes and styles to best suit the needs of musicians. Most MIDI controllers will feature a combination of performance features including keys, knobs, wheels, pads, joysticks, and sliders.

The orientation as well as number of individual performance features is decided by the manufacturer of the controller. As you probably already know, the keys on a MIDI controller are similar in look and function to traditional black and white piano keys. Knobs can either be continuous, meaning they can rotate continuously without stopping, or fixed point potentiometers, which have stops at both spectrums of the knob's turn. Like knobs, wheels come in two varieties as well, with one type being free to move while the other is spring loaded so it will bounce back to its starting position when released. Pads are rubberized squares that are touch and pressure sensitive. When using pads, the user physically smacks his or her fingers against each pad. As you can probably guess, a joystick on a MIDI controller looks and acts the same as a joystick on an arcade game. Finally, sliders are reminiscent of faders on a mixing console and act much in the same way.

Each MIDI controller will feature some of these types of performance features. The beauty of MIDI and software synthesizers is that a user can decide what each performance feature controls. This means that wheels can be assigned to pitch and modulation control while pads can trigger samples. Many creative setups can be created by assigning various parameters to features on a MIDI controller. Ultimately, it is up to each individual musician to create a setup that is both logical and comfortable while making sense for his or her particular needs. Most software synthesizers and even DAWs will have easy, user-intuitive ways in which to assign functions to a MIDI controller. One of the most common ways is having a MIDI message window where a user selects the particular function they wish to assign. Then, by turning the desired control on the MIDI controller while the function is highlighted, the software will automatically assign the highlighted function to that part of the controller that is being touched. A simpler version of that same method is the ability in some cases to right- or control-click a parameter, which will engage its assignment. When a parameter is engaged to be assigned, the user must then turn a knob or slider on his or her MIDI controller to assign the activated function. When assigning samples or individual drums to MIDI controller pads, the user will typically open an editor program that came with the MIDI controller and assign individual MIDI notes to each pad. Once MIDI notes have been assigned, the user can assign drum sounds or samples to be triggered via those MIDI notes inside of his or her DAW or software synthesizer. Ableton Live has really changed the climate for sample and MIDI controller interfacing. Ableton's intuitive clip launch features have made many software companies streamline their MIDI integration processes.

MIDI Controller Connections

When using sound modules, controllers will typically be connected to the module with a standard, five-pin MIDI cable. When connected to a laptop, a few options exist. The first option is to connect the controller with a MIDI cable into a fire wire or USB interface that is connected to the computer. Next, a MIDI-to-USB converter box can be in place, which will accept the five-pin MIDI cable from the controller and spit out a USB connection to the computer. Although these two setups are seen quite often, it is becoming even more common to have a MIDI controller that has a USB output on the back. In these instances, MIDI would be transmitted to the computer itself through MIDI over USB technology. With the introduction of Apple's Yosemite operating system, MIDI can now be transferred via Bluetooth. It is my prediction that fairly soon (in the next few months to a year), MIDI controllers will start shipping with Bluetooth built into them.

When connecting a MIDI controller to a tablet, there are fewer options. It used to be that the only way to get MIDI to input into an iPad was to buy the iPad camera kit that allowed for

Figure 9.12 Various forms of MIDI connections.

USB MIDI to be transferred. Now, dedicated devices exist to accomplish the same thing. Besides using MIDI over USB, many people choose to connect their controllers to their iPads via Wi-Fi. Although MIDI over Wi-Fi is an extremely cool tool, it oftentimes prove troublesome and, depending on your network, can even cut in and out frequently in short spans of time. Like with laptops, MIDI over Bluetooth is starting to gain more traction in the tablet world as well due to its stability.

One setup we have neglected to mention thus far is using a tablet as a controller for a software synthesizer installed on a laptop. To me, this is one of the strongest and creative setups available in the software world. By using an iPad as a MIDI controller, users are free to create a controller that is completely customized and suited to them and use it to control a synthesizer that may take up too much CPU power to be hosted on an iPad. As it stands now, there are far more software synthesizers available for computers than for tablets.

Many people also feel that the synthesizers that are available for tablets are not as powerful, nor professional, as their computer counterparts. Therefore, using a tablet as a customizable controller in which to control these more powerful synthesizers is hugely attractive. Also, tablets have both Wi-Fi and Bluetooth built in, allowing them to take full advantage of MIDI over Wi-Fi or MIDI over Bluetooth technology as it becomes more prominent.

Until now, we have treated the various platforms of synthesizer technology as entities that exist in a bubble. The beauty about synthesis is that you are free to use analog gear alongside digital gear as well as software in order to create your ideal setup. There is no wrong way to create a live, synth rig. Luckily, in today's day and age, there is technology available with which to interface all facets of synthesis technology while getting them to communicate with one another perfectly. For the first time in synthesizer history, not only is it possible to interface software with vintage analog gear, its actually quite easy. Being able to interface various technologies is just another step that helps musicians create their perfect live rig.

MIDI to C.V. Converters

As has been stated earlier, analog synthesizers both old and new use control voltage as a means of generating pitch and gate information. Since MIDI is doing this same thing, albeit with more options, it was only logical that someone would come along and create a way to interface the two technologies. A MIDI to C.V. converter or C.V. to MIDI converter is a device that filters MIDI information and converts MIDI pitch and gate information into voltages and vice versa. By using a MIDI to C.V. converter, artists can use digital sequencers, which are more complex than their analog counterparts, to control vintage or modern analog gear. MIDI to C.V. converters are perhaps the single best invention when it comes to interfacing modern and vintage gear.

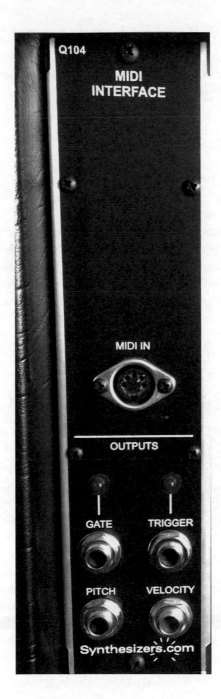

Figure 9.13 Synthesizers.com MIDI to C.V. converter module.

Volt/Octave Converters

As has been described, although most analog synthesizers adhere to the volt/octave standard control voltage, some synthesizers do not. A famous example is the Korg MS-20. The MS-20 was a hugely popular monophonic subtractive synthesizer introduced in the 1970s by Korg that instantly became a classic. The MS-20, however, uses a linear C.V. that is different from the volt/octave standard found on other synthesizers like Moogs. If you tried to control an MS-20 with a volt/octave synth or vice versa, you would soon find that the pitches were all over the place as well as the gate signal being reversed (meaning the note sounds when a key is released rather than pressed). With the new-found interest in analog synthesizers as well as Korg's reissue of the famous MS-20, companies have begun offering devices that can convert C.V. or MIDI to Korg's proprietary voltage levels. Most of these devices exist in Eurorack modular format and some work better than others. However, for the first time in history, an MS-20 can happily exist and be controlled with traditional volt/octave or MIDI synthesizers.

MOTU Volta

Although MIDI to C.V. converters allow for some software to analog communication, it is quite limited. When going from MIDI to C.V., typically the only thing transferred is pitch and gate information with a few exceptions in the form of trigger and velocity info. Parameters such as the filter cutoff frequency are not controllable via MIDI data in vintage analog gear. This is where MOTU's Volta software comes in. Volta is an instrument plug-in that effectively allows users to output control voltage from their audio interface. Many vintage analog synths feature control voltage inputs to various parameters for use with other synthesizers and controllers. Volta takes advantage of this by converting automation data written by the user into control voltage signals, which are outputted from the quarter-inch TRS outputs on an audio interface and then plugged directly into control voltage inputs on an

Figure 9.14 MOTU Volta
Software

analog synthesizer. Using this technology, it is finally possible to perform complex filter sweeps, envelope adjustments, pitch shifts and a variety of other techniques simultaneously without having to touch a knob on a synthesizer. When using Volta with vintage or modern analog modular synthesizers, the possibilities literally become endless. Using Volta live or in the studio opens up whole new worlds to synthesists and allows another level of software/analog compatibility.

Examples of Live Setups

At this point in a chapter, recipes are usually provided as means to not only convey the information discussed, but provide a jumping-off point for readers to begin experimenting with the information learned. As this live performance chapter is quite different from the individual synthesis format chapters, the recipes provided must be different as well. Therefore, this section will provide the reader with a number of possible live performance setups. These setups have been created outside of a historical context and are designed for the contemporary synthesist who has access to both vintage and modern equipment. It is highly recommended that each of these setups include a small, multichannel mixer so the synthesist can mix each synth or sound source together and provide the sound with a stereo feed. This keeps the individual sound levels in the control of the synthesist. By using a submixer, the synthesist is free to mute channels at his or her discretion to retune, adjust patches, or eliminate problem synths.

Setup 1: The Best of Both Worlds

The first setup in this list is designed with both analog and digital equipment in mind. This setup would bode well in an alternative band that might have some synth-heavy tracks as well as the need for traditional keyboard instruments and samples. The setup features a single centerpiece modern analog synthesizer as well as a multisound keyboard setup like a workstation synthesizer or MIDI controller and laptop. Finally, the setup includes a laptop computer with a sample-based synthesizer installed and a flexible MIDI controller. With this setup, the synthesist is free to utilize modern patch storage capabilities along with the huge, warm sound of analog synthesizers when playing lead, melody, or bass patches. By using a modern analog synthesizer, the synthesist can easily switch between patches in between songs. For songs that require a piano, organ, clav, or electric piano, the synthesist can quickly turn to their multisound keyboard. Finally,

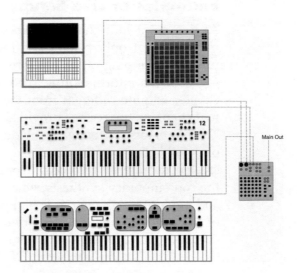

Figure 9.15 The Best of Both Worlds live setup.

when samples are needed, be it a piece of dialog, granular atmosphere, or song sample, the synthesist need only to trigger a sample with one of the pads on his or her MIDI controller. Although there are a number of individual instruments and devices that would fit these descriptions, the following is what we were imagining when creating this setup.

Modern Analog Synthesizer—Dave Smith Instruments Prophet 12
Multisound Keyboard—Nord Stage 2
Laptop—Apple Mac Book Pro
Software—Ableton Live
MIDI Controller—Ableton Push

Setup 2: The Jack of All Trades

This particular setup is designed for a scenario in which the synthesist is not only required to create inspiring melodies but also atmospheric and drum sounds as well. For this reason, this setup focuses on equipment that might be a bit out of

Figure 9.16 The Jack of All Trades live setup.

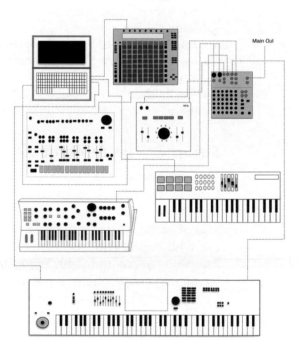

place for synth-only connoisseurs. This setup is centered on a laptop with a variety of virtual instruments being controlled by numerous MIDI controllers both keyed and padded. The setup also has a multisound keyboard with a bit more control and sculpting options than just a simple rompler unit. A powerful, yet compact analog synthesizer is also utilized for more sonic options. Finally, the setup is capped with a real-time, standalone audio manipulation unit. Using this setup, the synthesist is free to trigger loops and patterns via the laptop and MIDI controllers all while performing on the multisound keyboard and analog synth. At select points in the show, the synthesist can manipulate and warp sounds coming from the vocalist, guitarist, or any other musician in the band. The synthesist can create an entire environment for the music to exist in complete with pads, basses, leads, percussion, and samples. Again, there are number of instruments and controllers that fit the bill, but these pieces are what we were imagining.

Multisound Workstation—The New Korg Kronos Workstation
    Synthesizer
Laptop—Apple Mac Book Pro
Software—Ableton Live
MIDI Controllers—Ableton Push, Akai MPK 49
Optional Standalone Drum Machine—Roland Aira TR-8 or Akai
    Rhythm Wolf
Analog Synthesizer—Moog Sub Phatty
Audio Manipulation Unit—Roland Aira VT-3

Setup 3: The Millennial

This third setup focuses on harnessing the complete connectivity between mind, body, soul, and mobile devices with a helping of ironic perfectness. The setup is designed with the idea of creating a performance environment that lends itself to use with tablets and mobile devices while being able to exist in a physical environment. The setup is centered on a laptop, a tablet, a mobile phone, a powerful budget synthesizer, and a cheap, but surprisingly useful, toy synthesizer. The synthesist can trigger samples and patterns via a MIDI controller and laptop while having access to a variety of interesting

Figure 9.17 The Millennial live setup.

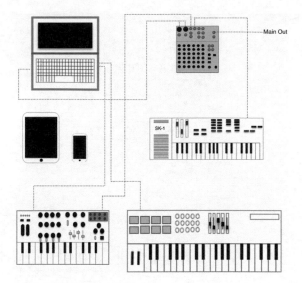

sounds on a tablet. The smartphone is used as an additional trigger source for use with the laptop. The budget synthesizer can be used to add additional bass or lead lines throughout the show. Finally, sounds can be sampled and played for live flair with what was once thought of as a toy. This is what we had in mind when creating this particular setup.

Laptop—Apple Mac Book Pro
Software—Ableton Live
MIDI controller—Akai MPK 49
Tablet—Apple iPad
Smartphone—Apple iPhone
Tablet Apps—Korg iMS-20, Korg Gadget, Moog Animoog
Phone Apps—Touch OSC
Budget Synthesizer—Arturia MicroBrute
Ironic Synthesizer—Casio SK-1

Setup 4: The Analog Purist

This fourth setup has the sound-creation purist in mind. This setup is for those who appreciate the sonic potential analog equipment offers without being swayed by the problems that may arise from bringing analog gear on stage. Since this setup is designed for the synthesist who craves the best of

Figure 9.18 The Analog Purist live setup.

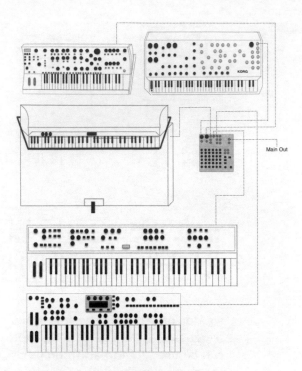

Main Out

the best, vintage and modern analog gear is used. The setup is centered on a powerful analog synth with sequencing capabilities as well as liberal I/O options. Next, a few select synths will compliment this centerpiece synth. No computers or tablets will be found in this setup: just pure analog goodness. The entire setup will be placed above or around a well-maintained, vintage electric piano. This setup bodes well for most styles of music ranging from ambient and alternative all the way to hip-hop and hard rock. The possibilities are endless for this setup. Here is the gear we had in mind.

Vintage Electric Piano—Fender Rhodes or Wurlitzer 200
Main Centerpiece Synth—Tom Oberheim Two Voice Pro Re-issue
　or Moog Sub 37
Polyphonic Synth—Sequential Circuits Prophet 5 or Roland
　Jupiter-8
Additional Synth—Vintage Korg MS-20
Additional Synth—Dave Smith Instruments Pro-2

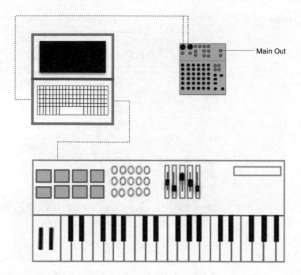

Figure 9.19 The Processor Programmer live setup.

Main Out

Setup 5: The Processor Programmer

This fifth setup is neat and tidy and relies on a good computer, controller, and user. The setup is designed to be run on a single computer with a single controller. The variety comes from the user's ability to map a number of sounds, functions, and performance-control capabilities onto different sections of the controller. Using this setup, the synthesist is free to have the keyboard divided into a number of sounds that all change for the next song with the press of a button. The equipment we had in mind for this particular set up is as follows.

Computer—Apple Mac Book Pro
Software—Ableton Live or Apple MainStage or FL Studio
MIDI Controller—Arturia KeyLab 88

Setup 6: The Wireless Wizard

This setup is designed around the desire to go completely cable free be it for logistical or merely cool factor purposes. The setup is rather small, but has a lot of potential for huge sounds. The setup is centered on a central hub laptop running Ableton live. Two tablets are then utilized for launching

Figure 9.20 The Wireless Wizard live setup.

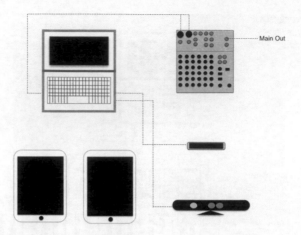

Main Out

clips while a body sensor senses limb movements and a hand sensor senses hand gestures. When using this setup, the synthesist will launch clips that can then be manipulated via the Xbox Kinect and Leap Motion sensors. Using a few programs for recognition and controlling, various gestures on the Kinect and Leap Motion can be used to control desired parameters inside of Ableton, such as delay feedback, reverb amounts, or even pitch and tempo changes. Here's the equipment we were imagining for this setup.

Computer—Apple Mac Book Pro (Yosemite or later operating system if MIDI over Bluetooth is desired)
Software—Ableton Live and Max for Live—Max/MSP
Tablets—Apple iPads
Body Sensor—Xbox Kinect
Hand Gesture Sensor—Leap Motion

Setup 7: The Eurorack Guru

The world of modern analog modular synthesizers is growing daily. There are a huge amount of companies producing analog modules for a variety of different size and power formats. By far the most popular is the Eurorack format. The modular company Doepfer, based out of Germany, helped bring about the monumental rise in Eurorack modular systems and helped solidify the format that has become known

Figure 9.21 The Eurorack Guru live setup.

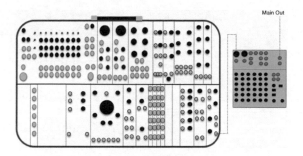

Main Out

as Eurorack. The Eurorack format is rather small and compact. The format relies on 1/8-inch patch cables and has a set height. The format most commonly uses +/-12V power with some major exceptions that require an additional +5V power rail. As the modular resurgence gains more popularity, more and more artists have begun bringing Eurorack systems to the stage. There are a wealth of difficulties that arise from bringing modular synthesizers to the stage. The most obvious is the complete and utter lack of patch storage. Due to the modular nature, it is almost impossible to re-create patches once knobs have been turned and patch cables disconnected. Changing a patch is also a nightmare as not only do parameters have to be adjusted, but the rat's nest of patch cables must be navigated and moved. This being said, bringing a modular synthesizer into a live environment is an exciting and rewarding venture. Due the large market of individual modules in the Eurorack format, we will provide an outline of a system that is capable to adapt to a variety of live performance environments and musical genres. Perhaps most important to consider when building a Eurorack system with the idea of performing live is the case. A case that can be closed and easily transported is a must, and the ability to close a case while leaving patch cables in place is greatly encouraged. Here's our ideal live Eurorack setup.

Case: Pittsburgh Modular Move [208] Double Row Case with internal power supply
Oscillators:
  ○ Pittsburgh Modular Oscillator (2).
  ○ Pittsburgh Modular Waveforms (1)

Filters:
- ○ Dave Smith Instruments Curtis Filter (1)
- ○ Modcan Multimode (1)
- ○ Grendel Formant Filter (1)

Envelope Generators:
- ○ Pittsburgh Modular ADSR (2)

Amplifiers:
- ○ Pittsburgh Modular Dual VCA (1)

Utility Modules:
- ○ Pittsburgh Modular Toolbox (1)
- ○ Division 6 Multiplicity xv (1)
- ○ Acidlab Mixer (2)
- ○ Intellijel Designs OR (2)
- ○ Intellijel Designs UM (2)

Modulation:
- ○ Pittsburgh Modular LFO2 (1)
- ○ Happy Nerding FM Aid (1)

Sequencer:
- ○ Division 6 Mattson sq816 Sequencer (1)

Effects:

- ○ Pittsburgh Modular Analog Delay (1)
- ○ Audio Damage aeverb (1)

Figure 9.22 The EDM Star live setup.

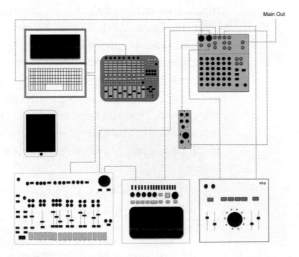

Setup 8: The EDM Star

The eighth setup in our list is geared towards electronic dance music festivals. This setup is designed to streamline signal and give the musician the ability to adapt the music based off of the responses of the crowd. The setup centers on a laptop and MIDI controller with a few hardware devices for processing and programming. Using this setup, the artist is free to play back entire sets with the ability to launch and mute clips, warp tempo and pitch, filter individual instruments or the track as a whole, and program bass and drum lines on the fly. This is the equipment we had in mind for this particular setup.

Computer: Apple Mac Book Pro
Software: Ableton Live
Main MIDI Controller: Akai APC40 MKII
Tablet: Apple iPad or Jazz Mutant Lemur
External Filter: Moog MiniMoog Voyager Rack Mount or stand-
  alone Moog 500 Series Filter
Hardware Drum Machine: Roland Aira TR-8 Rhythm Performer
Hardware Bass Machine: Roland Aira TB-3 Touch Bass Line
Hardware Processor: Roland Aira VT-3 Voice Transformer

Setup 9: The Retro DJ

At one point in synthesis history, when people imagined synthesizers onstage, they imagined DJs playing small clubs and warehouses. At this high time before the laptop's presence on stage, DJs used hardware instruments, samplers, drum machines, and mixers for making their music. Therefore, we have designed this setup with the hardware-inspired DJ in mind. For this particular setup, both vintage and modern gear will be examined. It should be noted that each of these setups is centered on an audio playback DJ mixer, such as vinyl or CD turntables in the vintage domain, and a laptop and mixing controller in the contemporary domain. The equipment listed below are the synthesizers and hardware devices needed to add all of the sounds, chirps, melodies, and bass drops necessary for the modern retro DJ.

Figure 9.23 The Retro DJ live setup.

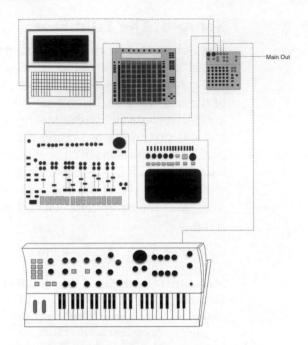

Vintage Setup

Drum machine: Roland TR-808 or TR-909
Bass Synth: Roland TB-303
Melody Synth: Moog Prodigy
Audio Playback Mixer: Vinyl Turntables and Mixer

Modern Setup

Drum Machine: Roland Aira TR-8 or Akai Rhythm Wolf, or New
  Korg Electribe
Bass Synth: Roland Aira TB-3 or Akai Rhythm Wolf
Melody Synth: Roland Aira System 1 or Moog Sub Phatty
Computer: Mac Book Pro
Software: Ableton Live
Controller: Ableton Push or modern MIDI mixer controller

Setup 10: The Art Instillation

The last setup focuses more on the limitless possibilities that synthesizers can offer the live performer. Synthesizers have always been, and will always be, harmoniously married with cutting-edge technology. It seems that with every advancement of technology and computational power, synthesizers are among the first things to benefit from the new technology. What is more, many musicians, engineers, and tinkerers are constantly inventing apps, software, controller interfaces, hardware devices, and various media-interfacing capabilities for synthesizers. A whole subculture known as controllerism has emerged that is filled with individuals designing complex fusions of art, technology, and music all centered on synthesizers and MIDI controllers. With the advent of programs like Ableton Live and Max/MSP, the creative climate for art installation synthesis is ripe. Although the possibilities are literally endless for the equipment one could use in this art installation movement, here are some options for getting started.

Hardware

Livid Instruments MIDI Brain
Arduino Boards
Rasberry Pi
Ableton Push
Escoteric Eurorack Modules
MIDI to C.V. converter

Software

Ableton Live
Max/MSP
Max for Live
MOTU Volta

Historical Perspective on the Synthesizer's Role in Live Performance

Despite synthesizers typically being considered studio tools, synthesizers have long been a staple in live performance. William Duddell was perhaps the first to incorporate an electronic musical instrument into performance when he wired multiple Oscillating Light Arcs together to play "God Save the Queen." Years later, the members of the Italian improvisational group Musica Elettronica Viva were pioneers in electronic music by becoming early adopters of the Moog synthesizer and realizing its benefits in a live performance setting. Using the huge Moog modular synthesizer live was no small feat. In the mid to late 1960s when the Moog system was beginning to get recognition, it was seen only as a permanent piece of hardware for large studios. In fact, Robert Moog often spoke about the burdens and impossibilities of lugging one of his systems to a show.[1] Aside from the sheer weight and size of a Moog modular system, the system itself was rather delicate and needed ample warm-up time and tuning to be remotely reliable. Add to this the time it takes to create a musical patch, and you can begin to see the difficulties inherent in gigging with a large, modular synthesizer. These concerns did not faze Musica Elettronica Viva, however, and they performed a number of improvisational electronic jazz pieces incorporating the Moog modular.

The next evolution of the synthesizer's role in live performance came during the infamous "Jazz in the Park" concert held at the Museum of Modern Art in New York City on August 28, 1969. Robert Moog and longtime collaborator Herb Deutsch were approached about using their newly created modular synthesizers for a live concert. Although skeptical at first, Moog and Deutsch finally agreed and decided to incorporate four individual Moog modular systems into the performance. The musicians included Deutsch on lead synthesizer, Hank Jones on Moog and electric piano, Artie Doolittle on modular bass synthesizer, and, finally, Jim Pirone on modular percussion synthesizer. The "Jazz in the Park" concert generated a large amount of interest for synthesizers and electronic music and helped sell a large number of Moog modular systems. Keith Emerson was rumored to have been present at the "Jazz in the Park" performance and ended up purchasing one of the Moog modular systems used that day.

Emerson would go on to have huge success with Emerson Lake and Palmer and is famous for lugging his large Moog modular onstage along with a variety of other synthesizers.[2]

The synthesizer's role in live performance would finally be brought to the mainstream with the introduction of the Mini-Moog Model D monophonic analog subtractive synthesizer. The MiniMoog finally brought synthesis technology to a small, portable format that could be adopted by musicians. The introduction of the MiniMoog created a paradigm shift in the music industry, and, soon, many other companies began offering small, portable synthesizers to meet the needs of touring musicians.

Throughout the 1970s, electronic instruments were used onstage mainly as additional instruments in rock, jazz, and fusion groups. At this point in electronic music history, synthesizers were mainly instruments set aside for traditional keyboardists, therefore, keyboard players would have a MiniMoog, ARP Odyssesy, or other small monophonic synthesizer stacked on top of their Hammond organs or Fender Rhodes electric pianos. During the second half of the 1970s, however, a class of musicians started to appear under the name of "synthesists." Frank Zappa and Devo were among the first to recognize the importance and promise of synthesizers and incorporated them as standard instruments in their repertoire.

Throughout the 1980s, many popular bands had designated synthesists who would create sequences, bass lines, lead lines, and even drum sounds on their synthesizers. Many synth pop bands in the likes of Depeche Mode, Gary Neumann, and Kraftwerk helped solidify the synthesizer's role in live performance. Once the synth pop genre became passé, the synthesizer went on to have an almost universal presence among the dance and rave music of the late 1980s and 1990s.

Synthesizers have gone on to appear onstage in almost every musical genre. The backing bands of pop icons can be seen hammering away on their synthesizers while the alternative rock super stars can be seen fiddling away on the knobs of their vintage analog synths. Hip-hop artists have long seen the benefit of synthesizers, and modern rock, blues, and even country performers have had the occasional synthesizer stacked on top of a piano or organ. One would be hard pressed to go to any show today and not see a synthesizer at least set up onstage.

Using synthesizers in a live setting can open up new, sonic worlds that might not have been imagined. Although playability and reliability might seem daunting at first, it is really no harder to incorporate a synthesizer into a live setup than any other instrument. Whether your particular setup involves gear that is strictly analog, digital, software, or a combination, many devices and programs exist to not only aid the performer, but to provide inspiring and helpful tricks for synthesis live performance.

Notes

1. *Moog*, DVD. Directed by Hans Fjellestad, 2004.
2. Thom Holmes, "Jazz Embraces the Moog Synthesizer," April 21, 2013. Retrieved from http://moogfoundation.org/moog-a-history-in-recordings-early-moog-jazz/

GLOSSARY

Ableton Live Versatile audio recording and live performance software environment.

AC *Alternating Current*; Oscillating electrical current.

Additive Synthesis A form of synthesis where sounds are created by combing single harmonics.

Alchemy A multiengine software synthesizer made by Camel Audio.

Algorithm In FM synthesis, an algorithm is the orientation of the onboard modulators and carriers.

Aliasing The introduction of unwanted frequencies in the digital to analog conversion process.

Amplifier In synthesis, an amplifier is a circuit that amplifies or attenuates sound when a gate signal is present.

Amplitude A measurement relating to the size of acoustical vibration; typically in reference to perceived loudness.

Amplitude Modulation Modulation of a synthesizer's amplifier.

AM Synthesis *Amplitude Modulation Synthesis*; A form of synthesis where sounds are created by the amplifier being modulated via an audio rate modulation source.

Analog Physical circuitry that utilizes voltages.

Analog Circuit Behavior Synthesis A form of analog modeling synthesis that attempts to model analog circuits and their behavior in an attempt to create a more realistic sounding re-creation.

Analog Modeling A type of physical modeling synthesis that aims to model the sound of analog synthesizers.

Arpeggiator A device often built into synthesizers that plays back notes being held down on the keyboard at a specified rate and order.

Artifact   Unwanted sonic material caused in the digital realm.

Asynchronous Granular Synthesis   A form of granular synthesis where individual grains are spaced randomly from one another in playback.

Attack   In envelope generators, the attack parameter controls the amount of time it takes a parameter to reach its user set amount once a key is depressed.

Band Pass Filter   A filter shape where frequencies both above and below a set band of frequencies are attenuated.

Bass   A descriptive term used to reference low frequencies.

Beat-Sync   The ability to align an LFO or clock source with tempo information; typically through MIDI beat clock.

Carrier   A signal that is modulated via a modulator.

Chorus   A delay type effect where each delayed sound is heard in rapid succession, causing the entire sound to be perceived as a single tone.

Clock   A repeating gate signal that is used to control sequencers, arpeggiators, and drum machines.

Combination Synthesis   A form of synthesis that utilizes two or more synthesis formats in its sound generation engine.

Complex Wave   Waves consisting of multiple overtones.

Control Voltage   A set of voltages that are used to control analog synthesizers.

Controller   A device that is used to generate pitch and gate information and relay it to a synthesizer; typically a keyboard, sequencer, touch plate, etc.

Cutoff Frequency   The frequency at which a filter begins its attenuation of the inputted sound.

Data Slider   A physical device that is used to control digital information.

Decay   In envelope generators, the decay parameter controls the amount of time it takes a parameter to fall back to its user set point after the attack time has run its course.

Decibel   A logarithmic measurement unit for sound.

Delay   An effect that takes a sound and reproduces it soon after its occurrence at a user set amplitude and rate.

Digital   Circuitry that uses integrated circuits and DSP.

Digitally Controlled Oscillator   An oscillator that is controlled via DSP instead of control voltage.

Driver   In physical modeling synthesis, a driver is in reference to how a sound is created; akin to a piano hammer striking a string.

Duophonic   The ability to play two notes simultaneously on a synthesizer.

Emphasis   See *Resonance*.

Envelope Generator   A circuit that receives a gate signal and allows users to set designations for how the sound will change in amplitude or timbre once a key is depressed, held down, and released.

Envelope Follower   A circuit which takes an audio input and creates an envelope shape that mimics the audio signal allowing for the synthesizer to be controlled by an audio device such as a voice or guitar.

Exciter   In physical modeling synthesis, an exciter is in reference to how a sound is created; akin to a piano hammer hitting a string. Another term for "Driver."

FFT   *Fast Fourier Transform*; An algorithm that is used to represent individual frequencies or harmonics present in a sound at any given time.

Filter   A device that attenuates frequencies in a sound.

Filter Modulation   Modulation of a synthesizer's filter cutoff frequency.

FL Studio   Digital audio workstation made by Image-Line.

FM Synthesis   A form of synthesis that generates sound by using audio rate oscillators, known as operators, to modulate other audio rate oscillators.

Formant   An emphasized frequency band; typically representative of speech sounds such as vowels.

Formant Filter   A series of band pass filters, placed in parallel, which are designed to mimic vocal formants and create vowel-type sounds.

Formant Synthesis   A complex form of synthesis typically used to create speech-type sounds.

Frequency   A representation of the number of vibrations per second in a sound; measured in Hz.

Frequency Modulation Modulation of a synthesizer's oscillator or tone generation source.

Frequency Response The measurement of electronic equipment's capabilities in regards to the frequency spectrum.

Gate A signal that is used as a trigger source; typically generated when a key is depressed. In analog synthesizers, a gate signal is made up of a +5volt signal.

Gating The technique of using a signal to prevent or allow another signal to be heard.

Glide A type of performance control that allows notes to gently rise or fall into each other; akin to a guitar string being bent from one note to another.

Glissando See *Glide*.

Glisson Synthesis A form of granular synthesis where each grain is modified by the presence of glissando.

Grain In granular synthesis, a grain is a multiple-millisecond-long fragment of sampled audio.

Grain Density In granular synthesis, grain density refers to how many grains are played per second.

Grain Duration In granular synthesis, the grain duration refers to how long a particular grain is; typically between ten and fifty milliseconds.

Grain Order In granular synthesis, grain order refers to the order in which the grains are played back.

Grain Playback Speed The speed at which individual grains are reproduced in granular synthesis.

Grain Stream In granular synthesis, grain streams are a designated group of grains.

Grain Window Individual grain envelope shape; typically limited to attack and release.

Grainlet Synthesis A form of sound synthesis.

Graintable A form of synthesis that melds granular and wavetable synthesis together for the software synthesizer Malstrom; made by PropellerHead.

Granular Synthesis A form of synthesis in which small fragments of samples audio, called grains, are played back at varying speeds, pitches, and orders.

Harmonics Overtones present in a sound that are multiples of the fundamental frequency.

Harmonic Series The integral order of harmonics present in a sound.

Headphones Device placed over or in the ears and produces sound.

High Pass Filter A filter shape where frequencies below a designated cutoff frequency are attenuated.

Hold In an envelope generator, the hold parameter places a set time between the attack parameter and the rest of the envelope generator parameters.

Hyper Wave A modified version of a traditional wave shape; i.e. *hyper-saw, hyper-square.*

Intensity The strength of a signal.

Joystick A stick-like controller which can move anywhere on an X and Y axis.

Key Follow A patch in which the higher one plays on the keyboard, the more the filter opens and vice versa.

Keyboard A pitch controller that features black and white keys in the same fashion as a piano or organ.

KS Synthesis *Karplus-Strong Synthesis*; A form of physical modeling synthesis that filters a short waveform through a delay line in an attempt to model hammered or plucked strings.

Leap Motion A controller that recognizes hand gesture and movement; can be used as a MIDI controller with third-party software.

LFO *Low Frequency Oscillator*; An oscillator that produces subaudible signals, which can be used to control various synthesizer parameters such as amplifiers, filters, and pitch.

Logic Pro A digital audio workstation software made by Apple.

Loom A software additive synthesizer made by AIR.

Loudness The interpretation of the intensity of sound.

Low Pass Filter A filter shape where frequencies above a designated cutoff frequency are attenuated.

MIDI   *Musical Instrument Digital Interface*; A protocol transmitted via 5-pin DIN and USB cables, which is used to control digital instruments and devices.

MIDI over Bluetooth   The act of transferring MIDI data over a bluetooth connection.

MIDI over Wi-Fi   The act of transferring MIDI data over a wireless connection.

Mixer   On a synthesizer, a mixer is a circuit that combines the various oscillators, noise sources, and external signals.

Modal Synthesis   A form of physical modeling synthesis that focuses on the frequency domain when modeling physical instruments.

Modular Synthesizer   A synthesizer where each component is independent of one another and must be connected together via patch cables.

Modulation   A routing in which a control signal is used to affect a component of a synthesizer.

Modulation Wheel   A wheel controller that designates the modulation amount.

Modulator   A signal that is used as a modulation source.

Monophonic   When only one note can be played at once.

MSW Synthesis   *McIntyre, Schumacher, and Woodhouse Synthesis*; A form of physical modeling synthesis that focuses on the time domain aspect of modeling physical instruments.

Noise Generator   A component that produces random noise, which can be mixed in with the oscillators.

Notch Filter   A filter shape that attenuates a small band of frequencies around a designated center frequency.

OSC   *Open Sound Control*; A control protocol that is transmitted via ethernet cable and can be transferred at faster speeds than MIDI information.

Oscillator   A tone-producing circuit on a synthesizer.

Oscillator Sync   The ability for an oscillator's waveform to be reset at the start of a separate oscillator's wave.

Overtones   Frequencies present in a sound that are above the fundamental frequency.

Paraphonic   The ability for a synthesizer to produce multiple notes simultaneously; each note is not an independent voice and shares envelope generators, filters, and amplifiers.

Partial   A harmonic added above the fundamental frequency in an additive synthesizer.

Patch   A sound created on a synthesizer.

Peak   See *Resonance*.

Period   The measurement of time it takes for a wave to repeat itself.

Phase Distortion Synthesis   A type of synthesis much like FM synthesis; typically uses a single modulator carrier pair instead of multiple modulator carrier pairs seen in FM synthesis.

Physical Modeling Synthesis   A form of synthesis that utilizes complex algorithms in order to mimic the physical properties of instruments in order to create a more realistic sounding re-creation.

Pink Noise   Noise that has equal energy at each octave interval.

Pitch Shifting   The ability to raise or lower the pitch of a sound.

Pitch Wheel   A wheel controller used to temporarily raise or lower the pitch of a synthesizer.

Polyphonic   The ability for a synthesizer to produce multiple notes simultaneously, each being complete, independent voices.

Portamento   See *Glide.*

Potentiometer   A variable resistor with a rotating or sliding contact; often referred to as *pot* or *knob*.

Preset   A stored patch on a synthesizer.

Pulsar Synthesis   A form of granular synthesis where each grain is created by an impulse generator.

Pulse Wave   A variable width waveform that contains all odd harmonics above its fundamental with a harmonic amplitude drop inversely proportionate to harmonic number.

Pulse Width   The width of a pulse wave in the negative and positive domains.

Pulse Width Modulation   The ability to vary the width of a pulse wave through a modulation source such as an LFO.

Quantize   The ability to shift MIDI notes previously recorded into proper alignment in regards to musical timing; in essence, a way to fix timing errors.

Ramp Wave (Reverse Sawtooth)   A waveform that contains the same harmonic content as a sawtooth wave, but with reversed shape; i.e., a short ramp up then instantaneous fall.

Release   In an envelope generator, the release parameter controls the amount of time it takes a parameter to fall back to zero once a key is released.

Resonance   A circuit that feeds the cutoff frequency back into the filter, effectively boosting the cutoff frequency and immediate surrounding frequencies, which causes the filter to ring.

Resonator   In Physical Modeling synthesis, a resonator is a control that adjusts the modeled resonance cavity or vibration source such as a drum shell or piano sound board.

Retrigger   The ability to start an LFO's wave cycle when a key is depressed independent of where the LFO is in its current cycle.

Reverb   The interpretation of a sound after it has been produced; resultant of reflected sound waves off of various surfaces.

Ribbon Controller   A controller that responds to touch in a sliding motion.

Ring Modulation   An effect that multiplies two signals together and outputs the sum and differences of each signal.

Sample   A previously recorded piece of audio.

Sample and Hold   A circuit that takes figurative snapshots of a voltage source at a set rate and then outputs those voltages; typically used as a random pitch generator.

Sample-Based Synthesis   A form of synthesis that utilizes recorded audio and then manipulates it through various synthesis techniques.

Sawtooth Wave   A waveform that contains all harmonics above its fundamental in the harmonic series with a harmonic amplitude drop rate, which is inversely proportionate to harmonic number; the shape, when viewed on an oscilloscope, resembles a saw blade with an instantaneous rise then sharp ramp down.

Sculpture   A physical modeling software synthesizer present in Apple's Logic Pro DAW.

Sequencer   A device that produces pitch and gate information and that is used to control a synthesizer.

Simple Wave   Waves consisting of limited overtones.

Sine Wave   A waveform that only contains a fundamental frequency.

Slew   See *Glide*.

Slider   A potentiometer that moves back and forth rather than in a circular motion.

Square Wave   A waveform that contains only odd harmonics above its fundamental with a harmonic amplitude drop that is inversely proportionate to harmonic number.

Sub Oscillator   A device that takes an oscillator, divides it, and then produces a tone that is one or two octaves below the original tone; once created, the new, lower, tone can be played along the original oscillator, creating the effect of two oscillators.

Subtractive Synthesis   A form of synthesis where timbres are created by removing harmonic content from a harmonically rich tone in both the amplitude and time domains.

Sustain   In an envelope generator, the sustain parameter is used to designate a level the sound will stay at, while a key is depressed, once the sound completes its attack and decay cycles.

Synchronous Granular Synthesis   A form of granular synthesis where individual grains are spaced equally apart.

Timbre   The way in which a sound is interpreted and distinguished; resultant of a variety of variables such as instrument makeup, force applied, overtones, and environment.

365

Time Stretching   The ability to elongate or shorten an audio clip without affecting pitch; typically accomplished through granular synthesis.

Touch Pad   A device that responds to touch on an X-Y axis and is used as a form of control.

Touch Plate   A device that responds to touch via pressure, heat, or internal body frequency, which is used to control a synthesizer.

Transistor   A device used to switch and amplify electronic signals; used heavily in analog synthesizer circuitry.

Tremelo   An effect where amplitude is quickly brought up and down creating a chop-type effect.

Triangle Wave   A waveform that contains all odd harmonics above its fundamental with a harmonic amplitude drop proportionate to the inverse square of harmonic number.

Trigger   A signal that is used to start a function such as a sequencer, LFO, or envelope generator.

Vector Plane   In vector synthesis, a vector plane is an arbitrary square that puts each sound source at each corner.

Vector Synthesis   A form of synthesis that contains four or more sound sources and moves the output from sound source to sound source, creating rich textures as the various sound sources mix.

Vibrato   An effect where pitch is slightly raised and lowered back and forth at a set rate.

Voltage   An electrical force; the creation of electrical current.

Voltage Controlled Oscillator   An oscillator that gets its pitch information via control voltage.

Waveform   The shape of a repeating wave.

Wave Folding   A technology that duplicates waveforms and then inverts and overlays them onto existing waveforms in an attempt to create richer textures.

Waveguide Synthesis   A form of physical modeling synthesis; a digital waveguide is a computational model for how waves propagate through physical media.

Wavelet Synthesis   Extremely similar to granular synthesis, but with stricter guidelines on grain length.

Wavetable   A compilation of the various waveforms available on a wavetable synthesizer listed in a particular order.

Wavetable Oscillator   An oscillator capable of creating a large number of wave shapes that can be swept continuously with little or no artifacts or sudden jumps.

Wavetable Synthesis   A form of synthesis where interesting textures are created by sweeping through a large number of wave shapes in order to create a constantly evolving harmonic atmosphere.

White Noise   Noise that has equal energy at each frequency.

Workstation Synthesizer   A physical synthesizer that contains a number of synthesis engines as well as recording and arranging capabilities.

Xbox Kinect   A motion sensor device manufactured by Microsoft; can be used as a MIDI controller with third party software.

INDEX